SOTHIS PRESS

Sothis Press aims to completely reimagine the public domain book with an emphasis on universal access to captivatingly designed print editions. We believe that technology will enable every single book ever published to be perpetually accessible in beautiful print form. Sothis is beauty in ubiquity.

www.sothis.press
1 Waterhouse Square, London, ECIN 2ST

# Harper's Aircraft Book

Why Aeroplanes Fly, How to Make Models,
and All About Aircraft, Little and Big

Alpheus Hyatt Verrill

SOTHIS PRESS

ISBN 978-1-0368-0175-5

*Harper's Aircraft Book*
is in the public domain.

This edition is published by
Sothis Press © 2023. The company
is registered in England and Wales.
The Company Number is 14900389.

# HARPER'S AIRCRAFT BOOK

WHY AEROPLANES FLY, HOW TO MAKE MODELS, AND ALL ABOUT AIRCRAFT, LITTLE AND BIG

BY

A. HYATT VERRILL

MEMBER OF THE TECHNICAL BOARD OF THE AERONAUTICAL SOCIETY OF AMERICA

AUTHOR OF

"HARPER'S BOOK FOR YOUNG NATURALISTS"

"HARPER'S WIRELESS BOOK"

FULLY ILLUSTRATED

HARPER & BROTHERS PUBLISHERS

NEW YORK AND LONDON

MCMXIII

PRINTED IN THE UNITED STATES OF AMERICA
PUBLISHED SEPTEMBER, 1913

I-N

# CONTENTS

## CONTENTS

# CONTENTS

## Part VI

### USES OF THE AEROPLANE

# INTRODUCTION

IN this book, which explains the making of model aeroplanes and the operation of large aircraft, the keynote is practicability, and the author's purpose has been to furnish a book on aircraft not only accurate, but simpler and more comprehensive than anything which he has been able to find on the subject.

In the preparation of the book he has had the aid and cooperation of many of the most noted and successful aviators in America, as well as of his fellow-members of the Aeronautical Society, the Aeronautical Bureau, Mr. Edward Durant, Mr. Hugo Rosenstein, Mr. Montague Palmer and other designers and builders of model aeroplanes.

The experiences, suggestions, and advice of such noted airmen as Capt. Thomas Baldwin, Harry Brown, Cecil Peoli, Frank Fitzsimmons, the late Howard Gill, and many others have proved of inestimable value, and have aided the author in substituting facts for theories, while details of construction, plans, and much other material of value have been freely given by many manufacturers, among them the Ideal Aeroplane & Supply Company, Thomas Brothers, Benoist Aeroplane Company, the Curtiss Aeroplane Company, Heinrich Brothers, the Sloane School of Aviation, and numerous aeroplane-motor manufacturers.

All the facts obtained from these authoritative sources have been combined by the author with his own personal knowledge to provide a book for boys which will be practical, up to date, and, he hopes, an advance upon anything hitherto prepared for a similar purpose.

The object of the book is twofold: first, to explain in a simple, lucid manner the principles and mechanisms involved in human flight; and, second, to tell boys how they may design and construct model aeroplanes, gliders, and even man-carrying machines.

To proceed in the design or construction of any mechanism without first mastering the whys and wherefores of the machine, the relation of one part to another, and the various influences and conditions which effect its successful operation, is putting the cart before the horse—a fault all too common in many books intended for boys. Before attempting to build even a model flier the builder should have some knowledge of the principles of flight, air pressures, speeds, lifts, and other details as well as the various mechanical details of aeroplane construction. As this book is intended primarily for youthful readers, a great many important but complicated and technical matters have been omitted, and the volume is not intended as a complete treatise on the modern airship, nor is it a history of aviation.

Only the most vital and salient features of construction and operation have been included in the chapters devoted to man-carrying machines, for, while it is quite within the power of any boys to construct a real and practical machine, yet it is more advisable for them to confine their efforts to model machines and gliders until thoroughly familiar with the principles and peculiarities of aeronautics, which can only be mastered by actual experience and numerous experiments, all of which may be readily carried out with model aircraft.

Many boys have already accomplished wonderful results

in the invention, design, and construction of model fliers, and the world's records for speed and distance flights of these miniature aircraft are held by American boys. It was in the winter of 1907 that the first model aeroplane club in America was organized in New York by Miss E. L. Todd. Through the efforts of the founder and Mr. Edward Durant the first model flight contest was held a little later in a New York armory, starting the models from the floor.

In December the first public exhibition was held in Madison Square Garden. This enlisted the interest of the West Side Young Men's Christian Association, which established a department of aviation. More and more attention was given to the new sport, larger buildings became necessary, and finally the aviators found it necessary to go out-of-doors to obtain sufficient space.

The contests of the young aviators have become a regular Saturday feature at some New York parks.

The Junior Aero Club of America was formed, and, with the quickening interest on the part of men as well as boys, the New York Model Aero Club was organized in 1910. Numerous thriving model aeroplane clubs are now in existence, and the boy members have regular meets, contests, exhibitions, and periodicals devoted exclusively to the model aeroplanes.

In this field of aeroplane construction there is ample opportunity for boys to obtain a great deal of pleasure, education, and practical knowledge, for the principles of flight in a small model are very similar to those of full-sized machines. It must not be supposed, however, that a model of a certain form or design will work equally well if enlarged to man-carrying size. For this reason the two fields of experiment and construction are very distinct, a fact that in many books on the subject has been overlooked entirely.

## INTRODUCTION

Gliders, although very similar in design to aeroplanes, are often efficient when aeroplanes of similar construction would fail entirely; but any boy can build a safe and satisfactory glider, and with these simple affairs may taste all the joy and exhilaration of actual flight without any of its attendant risk or danger.

Because model aeroplane and glider construction is work so admirably adapted to boys' abilities and needs, a great deal of space has been devoted to these subjects, and the author is confident that this book is unequaled in the amount of practical and detailed information in regard to aerial craft which it contains.

While the object of this volume is primarily to teach its readers how to construct model aeroplanes and gliders, yet those interested in real man-carrying machines, or in the progress or advance of aviation, will find herein a great deal of valuable information presented in such a simple manner that it may be readily grasped and understood.

To many the question of "Why does an aeroplane fly?" is of much greater moment than the method or manner of its construction, and many otherwise excellent books have been published which leave their readers as much in the dark regarding this matter as before. The author of this book believes that he has explained this puzzle so fully, so simply, and in such an elementary manner that any one can understand just why an aeroplane flies and the manner of its flight.

In the preparation of this work every effort has been made to avoid technical terms, involved or abstruse problems and formulæ, and to make all tables, explanations, and descriptions as simple, lucid, and concise as possible. Actual facts and results have been substituted for theories, and actual experiences of noted aviators and builders are given precedence wherever a question as to the most efficient design or construction occurs.

For this reason the descriptions of various types of machines and motors have been confined to thoroughly tried and proven forms in common use; and, while a vast number of efficient machines are in use to-day, yet all are merely variations of a few standard types, which if once mastered will make the others easy to understand.

In the majority of books treating of aerial craft and aviation—more especially those intended for boy readers—far too little attention has been devoted to the most important part of the aeroplane—namely, the motor.

The engine is the heart, soul, and life of the entire fabric, and upon the efficiency and reliability of this wonderful power the entire success of human flight depends.

The aeroplane motor represents the very highest development of the modern combustion engine; and, as motors of this type are so universally used in every-day life, every boy at all interested in mechanics should become thoroughly familiar with their principles and construction.

The chapter on motors in this book will be found complete, simple, and easily understood, for the motors selected for treatment are those of recognized merit and of distinct types.

No one can foresee the ultimate future of aviation, but in the minds of the author and many other leading authorities, the future flying-machine for commercial use will be of the hydroaeroplane type.

These machines, equally at home on water or land, or in the air, obviate many of the objectionable features of ordinary aircraft, and combine the advantages of the fast motor-boat with the power of sustained and rapid flight.

While any boy who can build a boat can also build an

hydroaeroplane, yet miniature or model "flying-boats" are better suited to the requirements and capabilities of the average boy. As such models are fairly accurate reproductions of the real hydroaeroplanes, very full descriptions of the principal types of these machines are provided, supplemented by directions for building models and describing the numerous details of parts, proportions, and materials required.

The illustrations are selected from hundreds of valuable photographs in the possession of the author, placed at his disposal by aviators and builders, supplemented by scale-plans and working-drawings, some original, and others loaned by manufacturers and by the publishers of *Aeronautics*, *Aircraft*, and other periodicals devoted to aviation. The drawings and plans have been prepared with the utmost simplicity possible combined with accuracy and detail, and are purposely more or less diagrammatic. To avoid the bugbear of computing measurements from scale, all such measurements have been plainly marked on the illustrations in addition to the detailed tables and specifications of dimensions included in the text descriptions.

Many boys are under the impression that an aeroplane, if only a model, is a complicated and intricate machine, difficult to build, handle, or understand. As a matter of fact, they are extremely simple affairs, far easier to master than an automobile, and safely handled with far less practice than is required to sail a boat properly.

Home-made, full-sized aeroplanes are, as a rule, far inferior to those built by professionals, and a boy who builds a real aeroplane should never attempt a flight, or even try to rise from the ground, until the machine has been thoroughly tested and approved by a competent aviator and the builder himself has become familiar with the operation and control of the aeroplane.

## INTRODUCTION

Many boys are ambitious to become aviators, and, while the author does not advise aviation as a profession, yet some boys become very expert and competent airmen. The only way to learn properly is to take regular lessons from a well-known and thoroughly competent pilot, numbers of whom make a specialty of teaching aviation.

By visiting the various aviation-fields and hangars a great deal may be learned that cannot be mastered by the perusal of any book, no matter how complete and intelligible it may be; and, as aviators and builders of flying-machines are invariably glad to point out and explain the various details of operation and construction of their machines to visitors, a great deal of time may be very profitably spent in this way.

# Part I

# WHY THE AEROPLANE FLIES

# HARPER'S AIRCRAFT BOOK

## Chapter I

### SAILING IN THE AIR

WE have all seen boats sailing against the wind, kites soaring steadily skyward in a brisk breeze, or have watched a hawk, buzzard, eagle, or other bird sailing in graceful curves or circles with no motion or effort of his wings. All these things are so common and so familiar to our eyes that we seldom wonder why they are possible or how the result is accomplished.

When at last success crowned man's attempts to rise and fly in a machine heavier than air, and aeroplanes rose and soared, twisted and turned, and performed marvelous evolutions in air, every one asked the question, "Why does an aeroplane fly?"

As a matter of fact, an aeroplane flies for precisely the same reason that a boat sails, a kite rises, or a bird soars; in other words, it flies because it *moves in the direction of least resistance, which is forward and upward.*

The boat, with its sail close-hauled, would merely drift sideways on the water but for the keel or centerboard, which offers a greater resistance to the water than does the stem, and hence, as the wind pushes sideways on the sail the boat moves forward. The simple kite cannot move straight away before the wind, as it is held by a string; and, as some relief must be obtained for the pressure against its surface, it slides edgewise. If the kite is not properly hung on its string or properly balanced by a tail or similar device, it dips and dives, performs wild gyrations and dashes to earth. If properly hung and balanced, the lower edge is farther from the wind's force than the upper edge, and the air, sliding from this slanting surface, forces the kite aloft, just as the boat moves forward and allows the wind to slide off the sail.

## Reaction

It can easily be seen that if, instead of a moving wind and a fixed surface, we substitute a fixed body of air and a moving surface the same result will obtain, and this is accomplished in the aeroplane.

A very important factor in the flight of a kite, the sailing of a boat, the success of an aeroplane, or the power of a windmill is *reaction.*

If a stream of water is turned against a flat surface at right angles it will spatter off in every direction. If the same flat surface is tilted, the water will rush off the object along the slanting surface.

If this flat body is so arranged that it can move edgewise but not directly away from the stream we will find that as soon as the surface is tipped, thus allowing the water to slide off, the surface will move forward *toward* the water (Fig. 1). This is caused by the reaction of the water; and, while the action of an air current cannot be seen as can the current of water, yet it acts in exactly the same manner. Thus any surface exposed to a moving mass or stream of air is acted upon in two ways: first, the direct pressure tending to lift it; and, second, the reactionary force which tends to drive it forward. In the windmill the latter is the only force utilized to obtain motion and power, but in kites and aeroplanes both forces are employed to advantage.

The amount of force or lift, as well as the amount of reaction, obtained depends upon the speed of the column of air striking the surface or the speed of the surface striking the air. A strong wind will drive a boat or windmill faster than a light breeze, but if we increase the size and area of the sail of the boat or the blades of the windmill without proportionately increasing the size of the boat's hull or the machinery of the mill nearly the same power and speed can be obtained with a light wind as the smaller sail or smaller mill developed with a hard wind.

As soon as we recognize this law of physics we can readily understand that speed and area are closely related, and that, while the earliest aeroplanes spread forty feet or more and traveled slowly, the latest machines have a speed of more than one hundred miles an hour and spread a scant eighteen feet.

### Surfaces or Planes

For a great many years men have realized that if a kite-like surface or plane could be driven with sufficient speed against the air it would lift and fly; but for years this was impossible to accomplish, as no known motive power could produce sufficient speed and power to drive a machine fast enough to lift its own weight.

In their experiments these men designed numerous machines and constructed many novel and wonderful forms of kites, for, lacking the power, they strove to evolve forms of planes or surfaces which would develop more lift. Such were the various "box-kites," the "Hargrave kites," "tet-

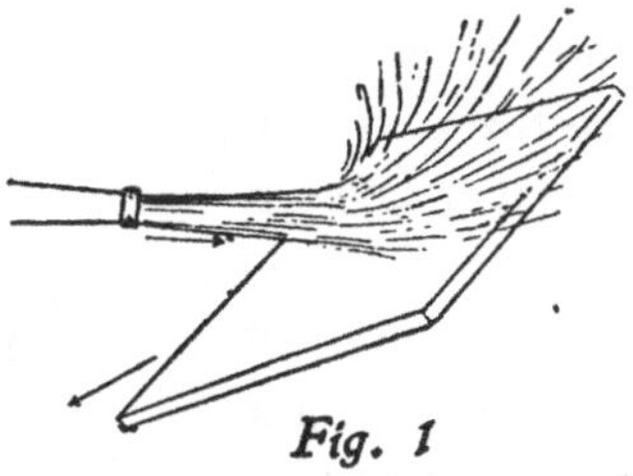

*Fig. 1*

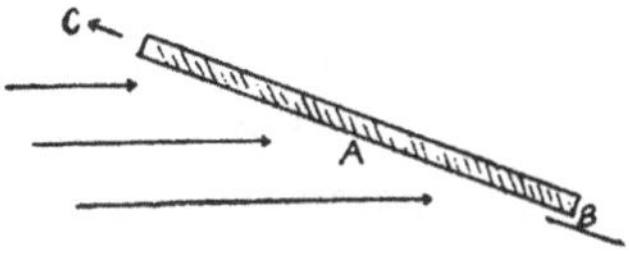


*Fig. 2*

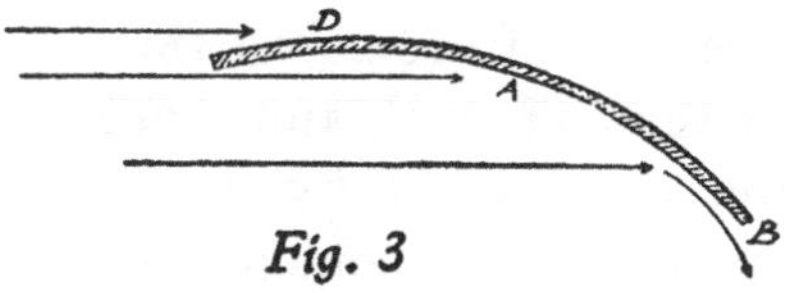


*Fig. 3*

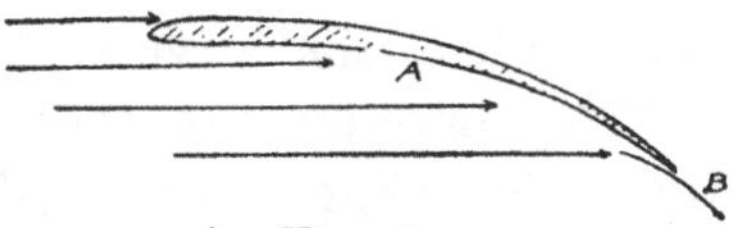


*Fig. 4*

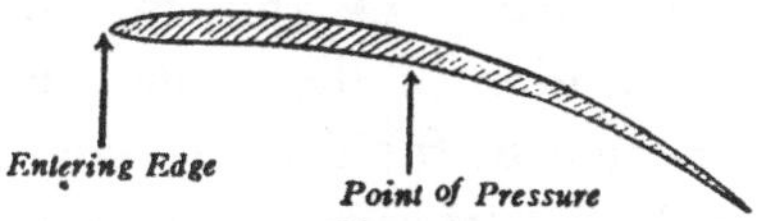


*Fig. 5*

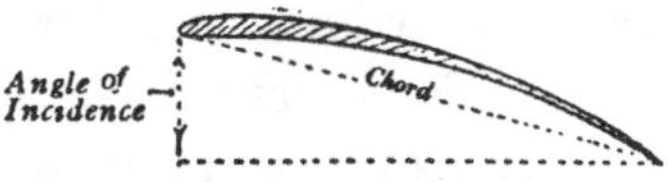


*Fig. 6*

rahedral kites," etc., some of which were so steady in flight, so great in lifting power, and of such large size that they carried men aloft with perfect ease and safety.

With the advent of the gasolene-engine, however, the long-sought power was found, and through this marvelous light, powerful engine which nowadays does such universal work for mankind the aeroplane became an actual fact.

The earlier wings or planes were perfectly flat, inclined surfaces much like an ordinary kite, but it was soon discovered that curved surfaces were far more efficient. Just why this is so can be easily seen by studying the diagrams in Figs. 2 and 3.

In Fig. 2 a flat, inclined surface is shown with the arrows indicating the direction of the wind. In this case the air strikes the flat surface and slides along it until it slips off the rear edge at A. The combined upward lift or pressure of the air at A and the reactionary force at B will, of course, force the entire plane toward C; but a large amount of the force is wasted owing to the fact that it is so very easy for the air to slip along the flat surface that only a small part of its lifting power is used. If a curved surface, as shown in Fig. 3, is used, we will see at once that the air striking the under side of the curve at A exerts a lifting power throughout the curve until it reacts or slips off at B, and that even when thus slipping off its direction, guided by the curve, tends still further to lift the plane. In a semicircular surface of this type the air approaching the surface at C must act more or less upon the upper surface at D, and this not only tends to retard the forward motion of the plane, but actually produces a downward motion of the forward edge. To overcome this fault "parabolic" curves are now generally used, such as shown in Fig. 4. Here the forward edge does not curve down enough to present the upper surface to the wind, while the long, gentle sweep of the lower surface allows the air to act upon it for its entire width.

This parabolic curve is now universally adopted in aeroplane and glider construction, but the exact proportions and curvatures vary with different designers.

The forward thickened edge of the wing is known as the "entering-edge," and upon its form and curve a great deal depends. The "point of pressure" is the particular point or spot upon the lower surface of the wing upon which the wind or air exerts its greatest force, and the distance from the entering-edge to the point of pressure has a very great influence upon the efficiency, stability, and success of any aeroplane, glider, or model (Fig. 5).

In every aeroplane the forward edge of the wings is slightly higher than the rear edge, or at an angle in relation to the horizontal line, and this angle is known as the "angle of incidence," which merely means the angle at which the wings meet the air as the machine moves forward (Fig. 6).

When a machine is once in the air it may be tipped at any angle required to rise upward, but as it rests on the ground and is run forward to gather speed the wings must present an angle in order to grip or receive the lift of the air currents. If the wings were perfectly horizontal the machine would merely travel on the earth and would never lift. This point should always be borne in mind when building gliders or

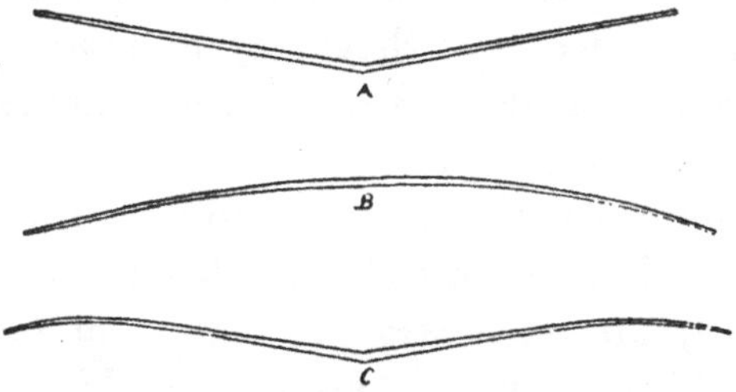


*Fig. 7*

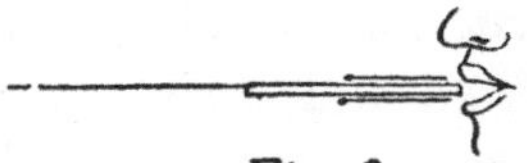

*Fig. 8*

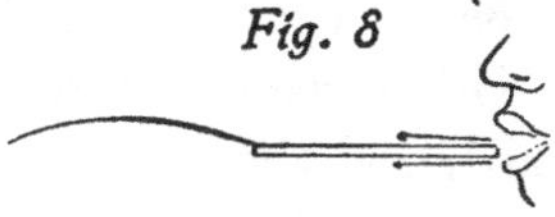

*Fig. 9*

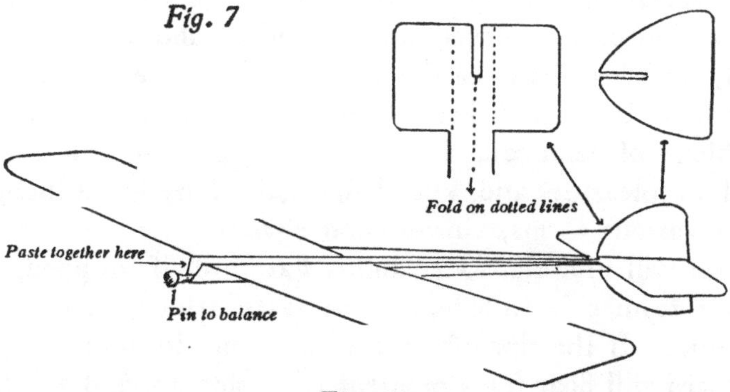


*Fig. 10*

models, and nearly every maker has a slightly different idea as to the best and most effective angle of incidence.

In addition to this angle the wings of most machines are placed at an angle or slope from their tips toward their center. This is the "dihedral angle" (Fig. 7, A), and it has been adopted because it gives greater stability or balance to the machine. In some machines the outer ends of the wings curve downward, producing "drooping wings" (Fig. 7, B), and such drooping tips in combination with the dihedral angle are often used with excellent results, although the droop alone lacks stability and balance. (Fig. 7, C.)

### Simple Experiments

These various points involved in wing form and construction are of vast importance, and they may be readily tested and their significance easily proved by very simple experiments with cardboard and paper.

If a piece of straight note-paper is attached to a strip of card, and the latter held before the mouth and blown upon, the action of the wind or air on a straight plane will be demonstrated (Fig. 8). If the paper is curved as in Fig. 9 the action of the curved surface will be shown.

By making little cardboard models of aeroplanes and balancing them with bits of wood, pins, etc., many of the problems of balance and flight may be grasped and a great deal of pleasure and knowledge gained by experimenting with various forms, curves, and shapes.

You will find that a straight, flat card, if dropped, will waver from side to side, and at last dash suddenly down edge on. If the tips of the card are bent down or drooped, the card will float for a moment right side up, and will then capsize and slide off at an angle bottom up. If, on the other hand, the card is bent sharply in the center so that the two halves are at an angle, the card will float quite steadily to the ground with little tendency to tip over or slide edgewise.

By attaching a body and tail to these cards, somewhat resembling the body and tail of an aeroplane (Fig. 10), you may so balance them, by trying various curves and weights, that when thrown by hand they will travel forward and even rise in the air for some distance.

We have now considered the action of the air or wind against the plane or wing, but we have overlooked one very important matter, which is the resistance of the moving body to the air through which it moves.

This resistance may be divided into true resistance and frictional resistance; and, as they are very important factors and exhibit some remarkable peculiarities, they should be well understood before any attempt is made to design or build any sort of an aircraft.

## Chapter II

### MOVING BODIES IN THE AIR

THE principal form of resistance produced by any body moving through the air is that known as "head resistance." This is the form of resistance which your body presents to a strong wind, and you all know how very hard it is to walk against a gale of wind. If the air was not moving at all, and you should run at a speed of forty miles an hour, you would find the resistance just as great as if you were standing still and the wind was blowing forty miles an hour. Fortunately, this form of resistance may be greatly reduced by using certain forms and shapes of bodies which offer a minimum amount of resistance to the air.

If a square body is exposed to wind or air currents (Fig. 1), the currents resulting will flow about as indicated by the arrows. If the body is spherical, as in Fig. 2, the flow will be more regular and "smoother," as shown by the arrows. In both of these forms the two separated currents do not meet and flow away behind until at some distance in the rear.

If this empty space or partial vacuum is filled in by a solid, the sides compressed, and the forward end sharpened, as shown in Fig. 3, the air-flow about it will be very smooth and gradual, or, in other words, this form of body will offer practically the least possible head resistance to the atmosphere. This is known as a "stream line form," or "fish-

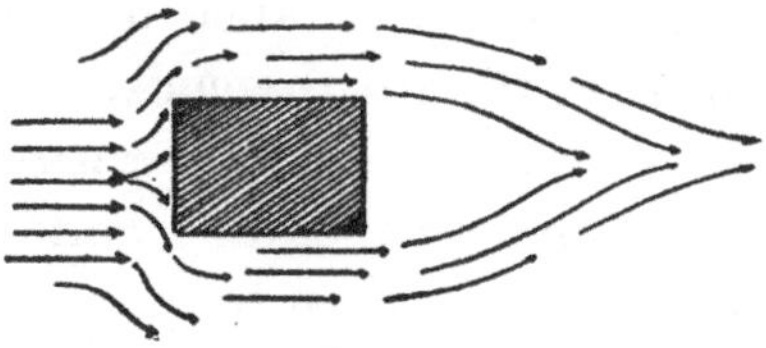

*Fig. 1*

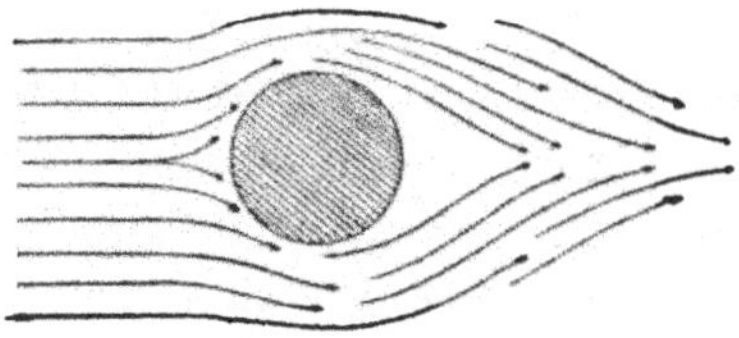

*Fig. 2*

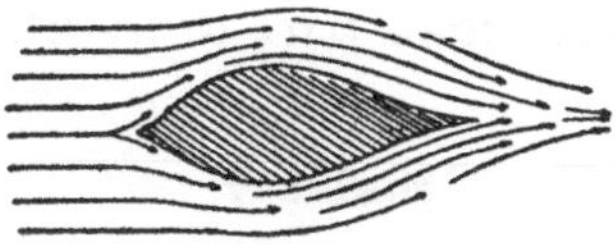

*Fig. 3*

body form," and is the sectional shape of most of the spars, struts, beams, and other parts of the modern aeroplane.

"Skin friction" is the friction or resistance caused by the moving air passing along and over the surface of an object. This does not amount to much on a perfectly smooth surface, but if the surface is rough or uneven a great deal of frictional resistance will result. For this reason every part of an aeroplane or other machine should be as smooth and even as possible, and even the surfaces or covering of the wings should be stretched tight and painted or varnished smooth.

As the greatest amount of skin friction occurs on the planes, these are made to offer the least possible friction with the greatest lifting area. The form which thus results is a long, narrow wing, as in Fig. 5, instead of a short, wide plane, as in Fig. 4. At first sight it seems as if the short, broad form would lift more than the other, but the arrows showing the direction in which the air slips off from the surface prove that the long, narrow plane has greater lifting power and less friction than the short, broad form. This is owing to the fact that the air will *always flow off from a body in the direction of least resistance.*

As the long, narrow plane presents less resistance to the air from front to rear than from front toward the ends, the air passes directly back, and in so doing exerts its entire lifting force. In the short, broad plane, on the other hand, the air finds less resistance toward the sides than from front to rear, and consequently slips off the sides without exerting more than a very small part of its lifting power on the plane.

### Stability

Another reason why the very long, narrow wings are used on aerial craft is because the short, broad wings are much less stable. Stability is a most important matter in aerial vehicles, and even under the best conditions only a very little real or natural stability can be obtained. The long planes serve to balance the body at their center, for their extremities act as levers or balancing-poles.

Just as a tight-rope walker maintains his equilibrium with a long pole, so the aviator finds the long wings of value in balancing his craft. Natural stability is the aim of every designer and builder of aeroplanes and models, and recent experiments have proved that certain remarkable phenomena are produced by bodies of peculiar form moving through the atmosphere.

One of these is the fact that any body moving through the air causes a disturbance of the air for a considerable distance *in advance* of the body itself. By numerous experiments and very complex formulæ students of aerodynamics have demonstrated that certain forms of bodies and wings will produce a disturbance ahead which will result in actually adding to the lifting power of the air. By adopting wings designed to take advantage of this peculiar fact very much greater efficiency has resulted.

Another discovery which has quite recently been made is the fact that, whereas the old-fashioned curve of wing and wing-beams tended to reduce stability when ascending or descending, more deeply curved wings and thicker and more rounded entering-edges result in increased stability. This seeming paradox is the result of the "center of pressure" changing as a machine rises or descends, and whether this point changes to front or back depends largely upon the curve of the wing section.

The diagrams Figs. 6 and 7 make this matter much clearer than an explanation. In Fig. 6 a conventional form of monoplane is illustrated. A shows the machine flying in a horizontal plane, with the center of pressure indicated by the arrow; B shows the same machine rising, and here it will be noticed that the center of pressure has moved forward, thus tending to tilt the machine still farther up with possibly fatal results; C shows the same machine descending, and here the arrow indicates that the center of pressure has moved to the rear, thus tending to tip the machine-tail up and cause it to "dive."

In Fig. 7 a monoplane of the same form with improved wings is shown. A illustrates the machine in horizontal flight, with the center of pressure indicated by the arrow. If this machine ascends, as in B, the center of pressure, instead of moving *forward*, moves *back*, thus tending to preserve balance or stability. At C this same machine is shown descending, and here again the center of pressure moves, but it moves *forward*, thus helping to preserve stability and preventing the machine from diving.

In making your models or gliders, or in designing a large machine, you should always bear these facts in mind and try numerous experiments with wings of various forms and curves before finally deciding on the one you will adopt.

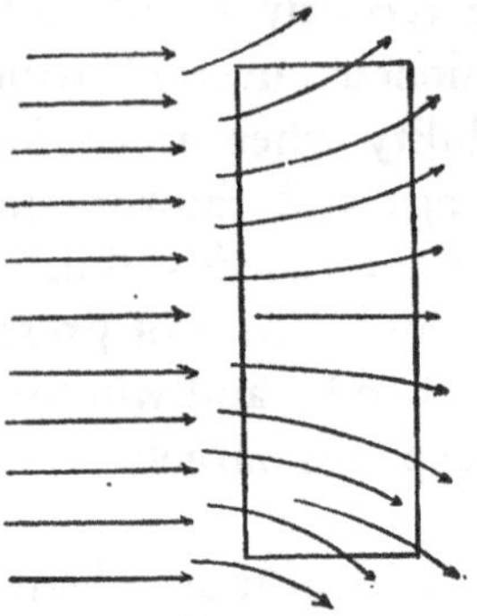

*Fig. 4*

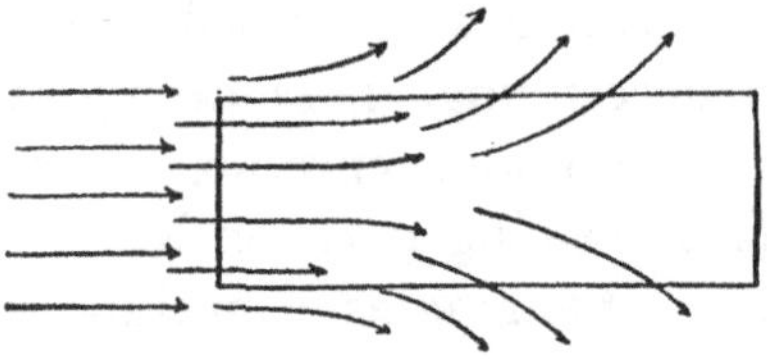

*Fig. 5*

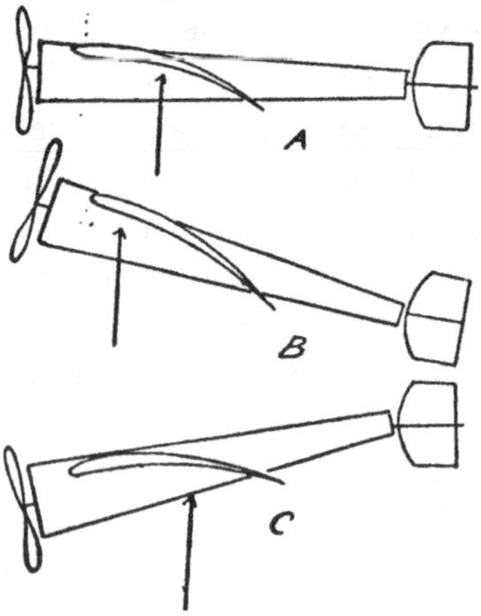


*Fig. 6*

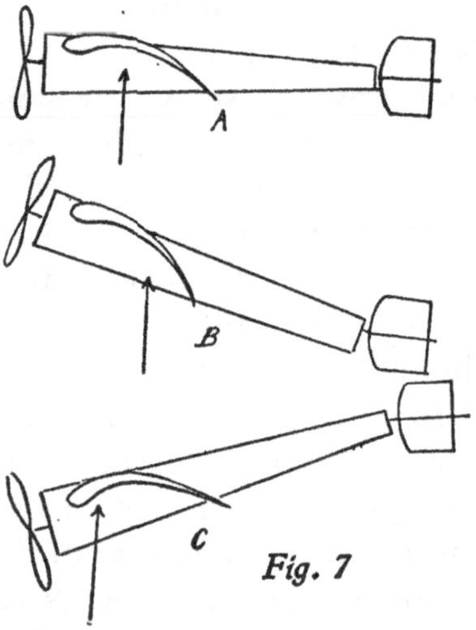


*Fig. 7*

No machine has been made which is entirely dependent on natural or "inherent" stability, for, while small models can be so constructed, in man-carrying sizes they are uncertain and liable to capsize. Moreover, the small machines are not intended for flying in circles or performing figure eights, spirals, etc., which are necessary in real flying-machines.

The early machines could not turn at all, and it was a

long time after the first flights were made in public before the aviators learned to turn a corner, or even a half-circle, in the air.

Each turn, twist, or deviation from a straight course results in a variation in balance and stability, and to overcome the constantly changing equilibrium, as well as to allow the pilot to turn or manœuver in any direction, various supplementary or auxiliary devices are used. Some of these are used in turning to right or left, others for rising or descending, and still others for maintaining balance or stability. All these things must be under the direct control of the operator, and must be easily and quickly reached and manipulated, for upon them his life invariably depends.

These various devices are collectively known as "controls," and as they vary a great deal in several types of machines, and, as they are in a way the most important parts of the aeroplane, they should be carefully studied and thoroughly understood before attempting to make or fly any sort of aeroplane or glider.

Very few model aeroplanes are provided with controls; but it is great fun to fit them to models and, by setting them in various ways, watch the effects they have upon the flight of your tiny machines.

As those used on model machines are identical in action and design with those used on real machines, a knowledge of one will answer for both, and in the next chapter I will describe and explain the most important controls and their most noteworthy and efficient types.

## Chapter III

### STEERING IN THE AIR

THE main control of any aeroplane is that which renders it possible to guide or steer the machine upward when rising from the ground, or to steer or guide it downward when the pilot wishes to descend or "volplane."

This control may be obtained by horizontal rudders in front of the machine, or by a horizontal tail in the rear, or a combination of both. In the older forms of machines, such as the original Wright, the old Curtiss, the Farman, and others, the front horizontal rudders, or "elevators," were

*Fig. 1*

always used. Nowadays the tendency is to omit the "head" and depend entirely upon rear elevators. Fig. 1 shows an old-style Farman machine with the forward elevator, or head, whereas Fig. 2 represents the Burgess-Wright biplane without the head.

The action of either the forward or the rear horizontal rudder is very simple and easily understood. The rudders

*Fig. 2*

consist merely of small planes of fabric stretched on a wooden frame, and pivoted or hinged so that they can be swung up or down like a trap-door or the lid of a chest. If the rear elevator is moved as shown in Fig. 3 A, or the forward elevator is swung as in Fig. 3 B, the pressure of the air against the angle of the elevator forces the machine to tip upward and ascend as shown by the arrows in the cut. If the pilot wishes to descend, the elevating-planes are swung in the opposite direction, and the forward end of the machine will be forced down by the pressure on the elevators, and the entire aeroplane will follow the direction shown by the arrow.

Monoplanes always have the elevating-planes in the rear, and, while they may vary in size, position, or details of construction, their principle and action are always the same.

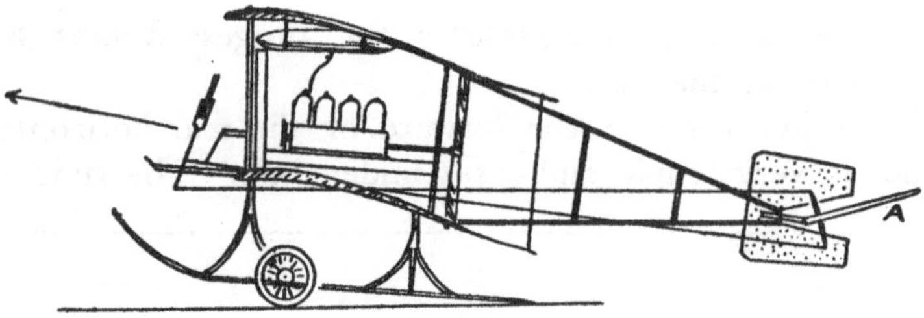


*Fig. 3 A*

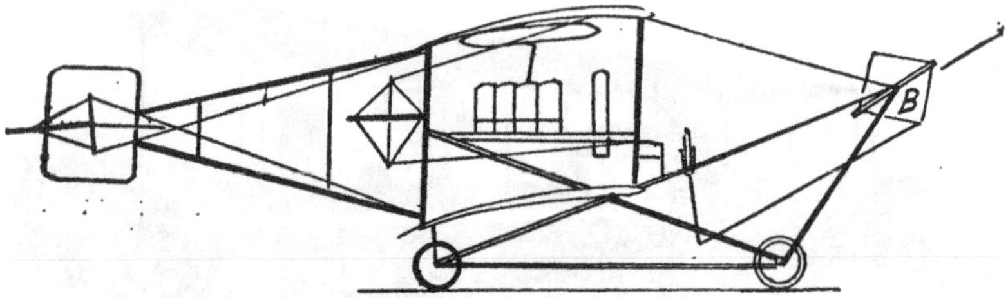


*Fig. 3 B*

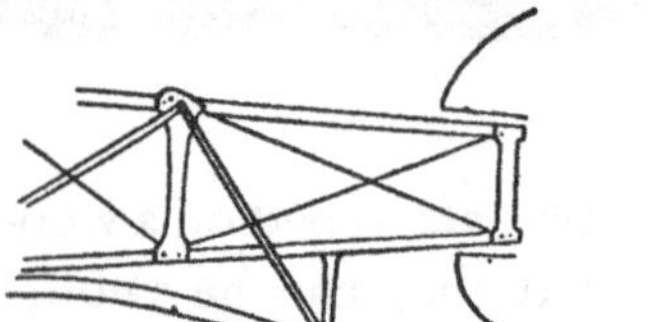


*Fig. 4 A*

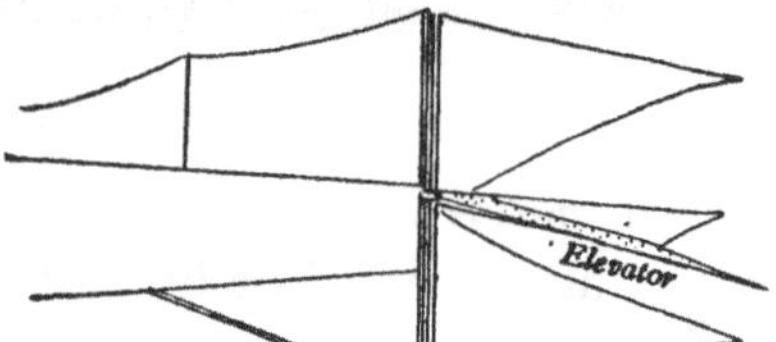


*Fig. 4 B*

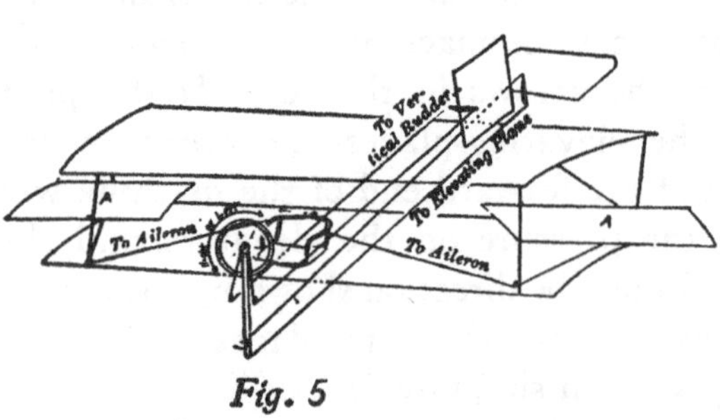


*Fig. 5*

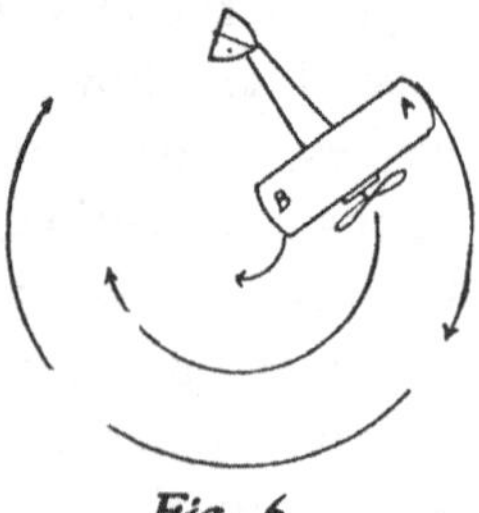


*Fig. 6*

HOW TO STEER AN AEROPLANE

The elevating-planes of a Bleriot monoplane are shown in Fig. 4 A, and those of an Antoinette in Fig. 4 B. The manner in which the elevators are moved or controlled by the aviator varies greatly in different machines, but in the majority of biplanes the elevators are connected by wire cable to a steering-column, or post, and by moving this forward or back the pilot raises or depresses the elevators (Fig. 5). Only a very slight motion of the horizontal rudder is required to cause the machine to vary its angle of ascent or descent a great deal, and in building model aeroplanes you will find that the least variation of the angle of the planes will make a wonderful difference in their flight.

For a long time after aeroplanes flew successfully it was impossible for the operators to turn in a curve or circle without upsetting. The tails of the machines were arranged to move like rudders to boats; but, nevertheless, a slight turn would upset the balance of the machine and cause disaster. It was not until the Wright brothers invented the "warping-wings" and other inventors discovered "ailerons" that stability could be maintained and circles, figure eights, and other evolutions performed in midair.

If we examine the diagram Fig. 6, which represents an aeroplane turning in a circle, we will find that the outer wing (A) must travel a great deal farther than the inner wing (B), just as the outer rim of a wheel turns farther in making a revolution than does the hub. In order to accomplish this the outer tip of A must travel through the air at a much greater speed than the inner wing-tip (B). As I have already explained that the faster a plane moves the more lift it produces, it will be easily understood that the excess in speed of A must tend to lift the outer wing-tip and that side of the aeroplane a great deal more than the slower-moving inner wing.

Of course, this causes the machine to tip more and more toward the center, and if carried too far would cause the aeroplane to slide sideways to earth or to capsize. In order to overcome this tipping, or "banking," and enable the pilot to turn safely, various devices known as "warping" or flexible wings and ailerons were designed.

*Fig. 7*

The warping-wing consists of a flexible portion of the outer end of the planes connected by cables with the operator's control in such a way that the wing or plane may be bent or twisted (Fig. 7). By bending down one wing and bending up the other the planes can be made to present a greater or less resistance and lift to the air. Thus, in turning a corner, as shown in Fig. 6, the speed of the outer wing, A, may be decreased by warping down that wing, and the speed of B accelerated and its lift increased by the same control-cables, so that the tendency to tip up is neutralized and the machine remains practically on a level keel.

The same result is accomplished by means of ailerons in other machines. These are merely small auxiliary wings or planes which are placed either at the rear edge of the main planes or midway between them, as in Fig. 8, AA. The function of the ailerons is exactly similar to that of the warping-wings, for by tipping them up or down one or the other wing-tip may be retarded and its lift decreased. Ailerons and warping-wings are known as "lateral controls," and are used in maintaining side balance or "lateral stability." In addition to their value in turning corners, they are also of the greatest importance in aiding the operator to keep his machine on an even course when flying in disturbed air, in uncertain winds, or, in fact, wherever it is necessary to keep the machine from tilting first one way and then another.

Each kind of a machine has a distinctive method of operating the cables leading to the warping-wings or ailerons. In the Wright machines it is accomplished by a hand-lever (Fig. 9 A); in the Curtiss it is brought about through a shoulder-yoke (Fig. 9 B); in the Bleriot it is done by tipping the steering-column (Fig. 9 C); and in other makes it is by foot-pedals or steering-wheels. Most operators prefer the shoulder-yoke, for in this system the natural tendency of the pilot to swing his body away from the side toward which his machine tips operates the aileron control and automatically rights the aeroplane.

If you watch an aeroplane carefully while in flight, you will notice that the ailerons are continually moving ever so slightly, and in many biograph pictures this may be seen to splendid advantage. When warping-wings are used they do not show their motion so much, but they are just as constantly in operation, and the pilot depends upon them just as much as in the case of ailerons.

Fig. 8

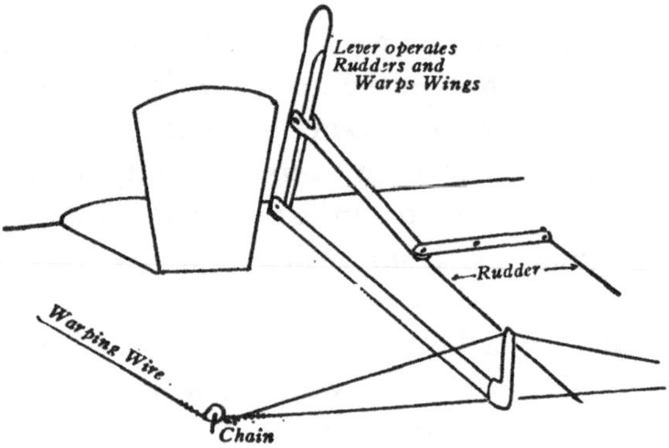


Fig. 9 A

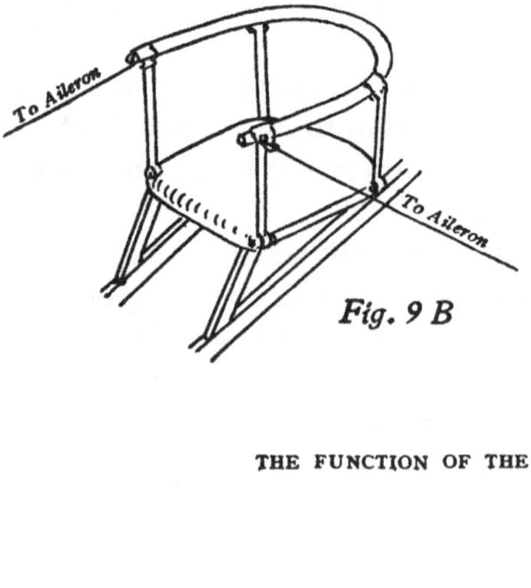


Fig. 9 B

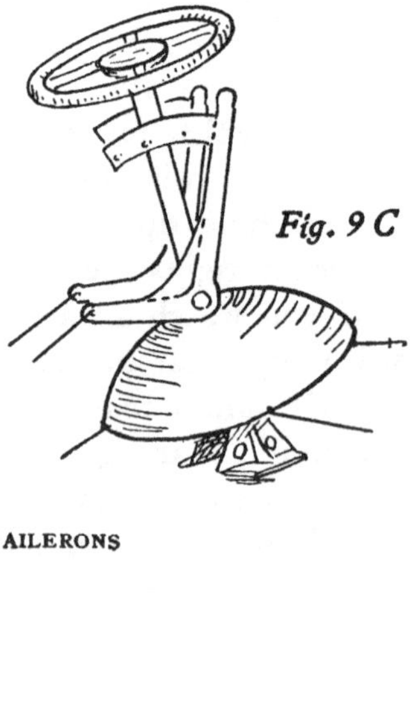

Fig. 9 C

THE FUNCTION OF THE AILERONS

Motion from one side to the other, or turning, is accomplished in the air very much in the same manner as a boat is turned in the water—namely, by a vertical rudder. The rudder of an aeroplane is merely a light frame covered with

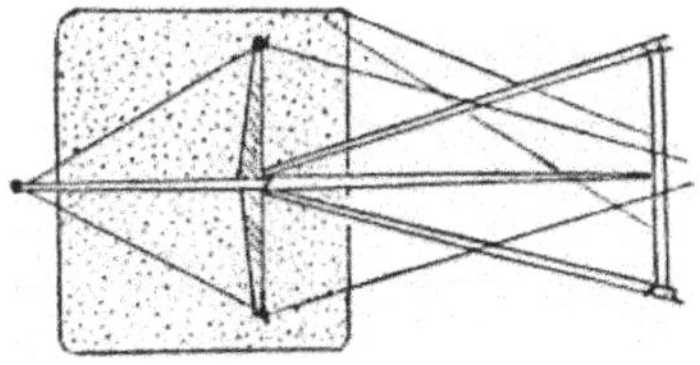

*Fig. 10 A*

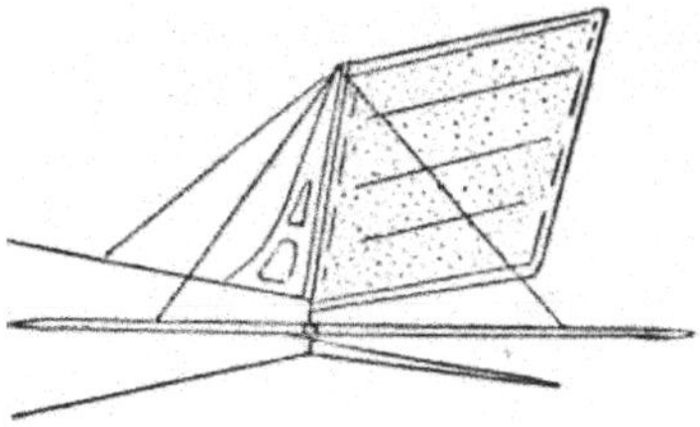

*Fig. 10 B*

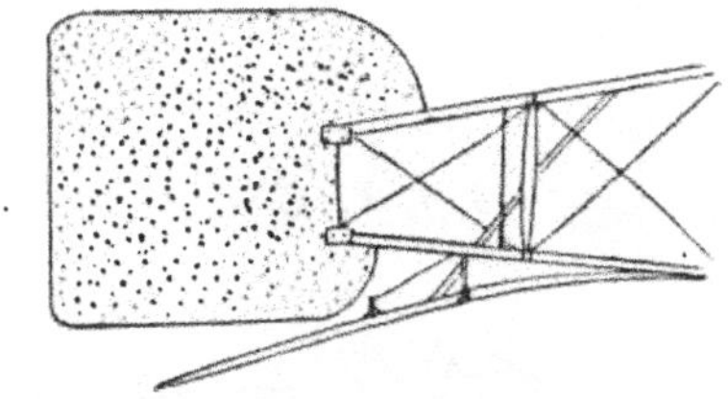

*Fig. 10 C*

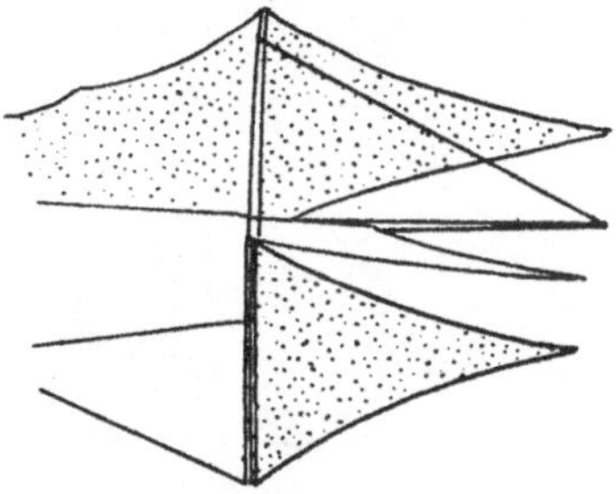

*Fig. 10 D*

fabric and usually placed in the rear of the machine on the tail, with about half of its surface above and half below the horizontal tail and elevators. Fig. 10 A shows the rudder on the tail of a Curtiss biplane, Fig. 10 B the rudder on the tail of a Benoist, Fig. 10 C that of a Bleriot monoplane, and Fig. 10 D the rudder and elevator of an Antoinette.

The size, position, and details of construction vary with each designer and type of machine, and in some cases the vertical rudder is placed in front of the machine in the head with the elevators, or forward and rear rudders are both used, each working in unison with the other.

In many forms of machines the vertical or steering

rudder is so arranged as to operate in unison with the warping-controls or with the ailerons in such a way that when the machine is turned to right or left the proper aileron or wing-tip is raised or lowered, thus more or less automatically maintaining stability.

Several typical forms of tails, rudders, and ailerons are illustrated in Figs. 11 and 12. Fig. 11 A shows the new

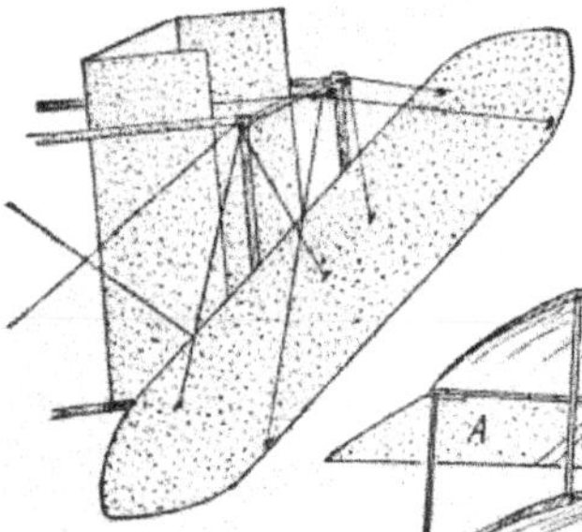

*Fig. 11 A*

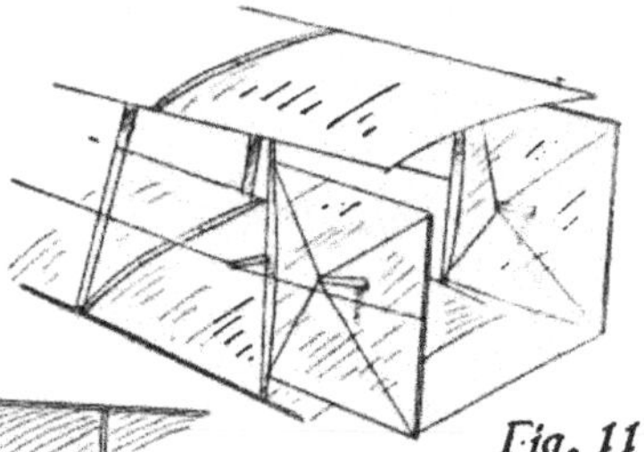

*Fig. 11 B*

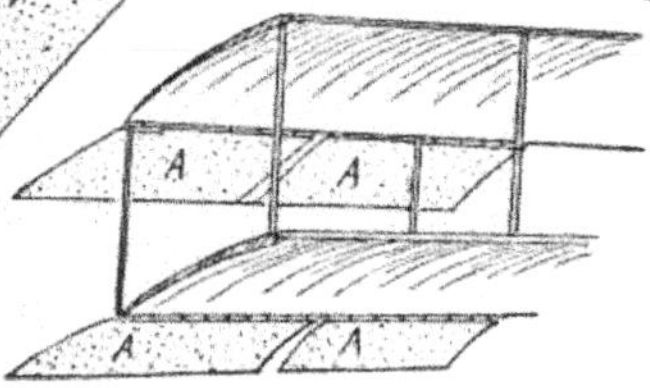


*Fig. 11 C*

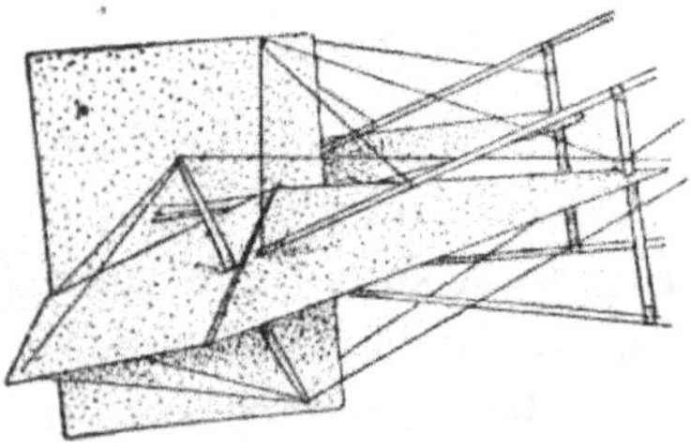

*Fig. 11 D*

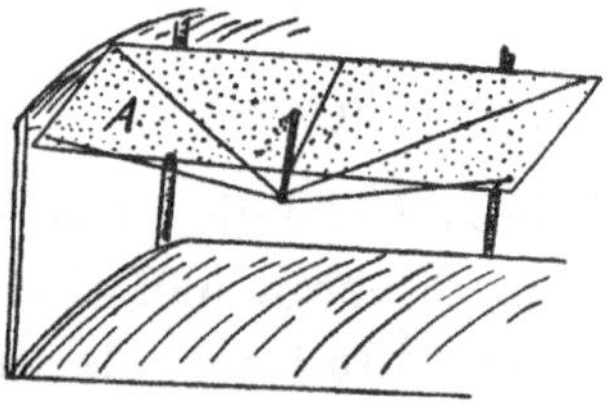


*Fig. 11 E*

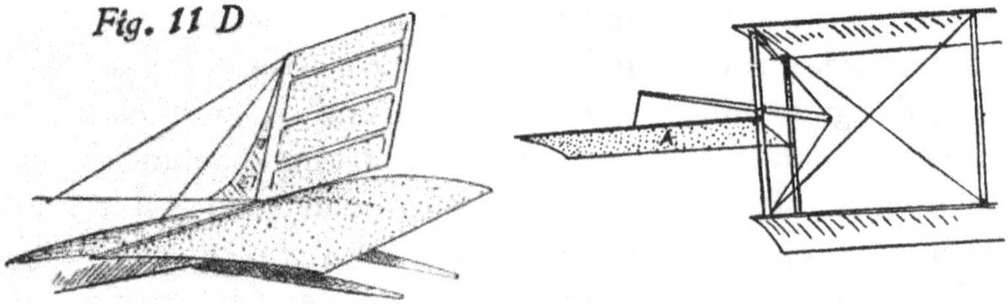


*Fig. 11 F*

Wright form of vertical and horizontal rudders; B is the Voisin form used on the Farman aeroplanes in conjunction with ailerons attached to the rear edges of the main planes, as shown in C; D is the famous Curtiss tail and rudders used with the ailerons between the wings shown in E. The tail and rudders of the Benoist machine, and its odd ailerons extending beyond the wing-tips, are shown at F. These are all typical biplane devices, and several monoplane rudders are shown in Fig. 12. As monoplanes use warping or flexible wing-tips, and not ailerons, the tails and rudders alone are illustrated (Fig. 12). A shows the Bleriot, B the Heinrich, C the Columbia.

Any of these forms or their modifications are capable of

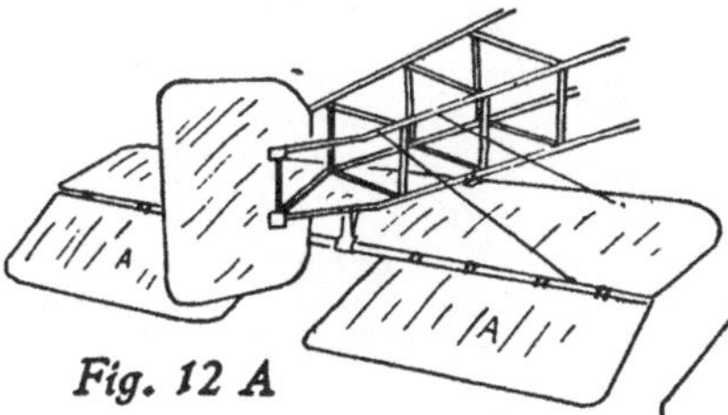


*Fig. 12 A*

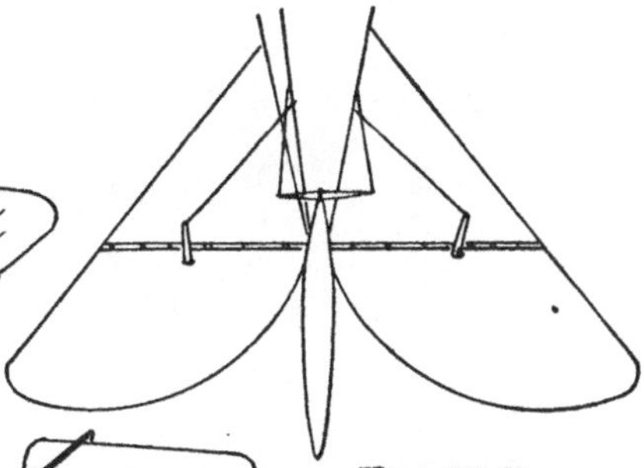

*Fig. 12 B*

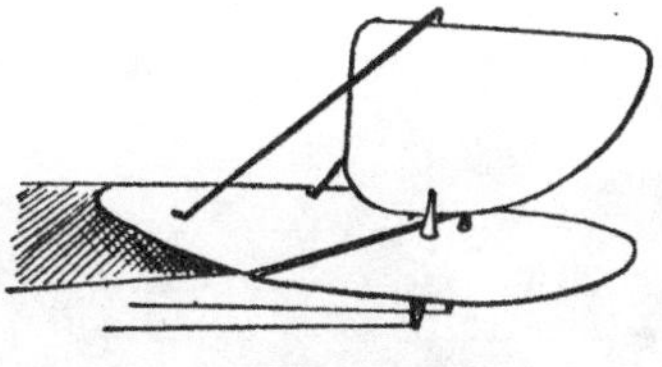

*Fig. 12 C*

adaptation to model machines, and there is really no particular choice as regards their relative efficiency.

No matter what system of rudders, elevators, and lateral controls are used, all three *must* be present in some form in any real aeroplane, even if it is of the peculiar tailless type made by the late Frank Boland (Fig. 13), or the tail-less Voisin-Canard machines. In the Boland aeroplanes lateral control is effected by "jibs" between the planes, and the

*Fig. 13*

elevating-rudders are ahead of the machine on an outrigger.

In model fliers the controls may be greatly simplified, for lateral stability is of little importance, as these miniature machines are intended mainly for straightaway flights. Moreover, for straight flights vertical rudders are unnecessary, and the elevation is secured by the planes or wings being set at fixed angles.

Sometimes excellent results are obtained by upright fins or vertical planes fastened across the main planes. These act like the keel of a boat, and cause the machine to fly very straight and true.

In designing or building any machine, whether man-carrying or model, it must be borne in mind that all the horizontal attachments, such as elevators, ailerons, etc., add to the total area of surface and lifting power while vertical controls add to skin friction or resistance without any additional lifting force.

# Part II

# MODEL AEROPLANES AND FLIERS

## Chapter IV

### MODEL MACHINES

THE term "model aeroplane" is in a way very misleading, for a true model aeroplane is an exact copy in miniature of a real plane. Such a model may or may not be capable of actual flight, and some of the most beautifully made models of this class, with every detail worked out with utmost care and accuracy, can never be flown. Such models serve only to illustrate the proportions, construction, and appearance of the real machines, and have little interest save as curiosities and examples of good handiwork.

Actual flying-models of large machines are often seen, and some of these are marvels of good work, accurate attention to details, and are most interesting and instructive, for they illustrate far better than any other method the actual principles and operation of the big machines.

Nowadays the term "model aeroplane" is more widely applied to small machines which are very distinct from any man-carrying aeroplane, and are designed and built solely for amusement or for racing or distance contests.

These machines are more properly known as "racers" and "fliers," and it is with these that the various "model aeroplane clubs" hold contests, and for which prizes are offered. Such models are of no real value as related to large machines, for, oddly enough, a flier that works well and breaks records would fail utterly if enlarged to man-carrying dimensions.

If a boy wishes seriously to study aviation with a view to ultimately designing or building a real aeroplane, much greater progress may be made and much more knowledge gained by building small reproductions of standard machines than by constructing fliers and racers. On the other hand, a great deal more amusement and real fun may be obtained from the fliers than from true models. Fliers are cheap and easy to build, and one never knows when some new or untried design may prove to possess remarkable powers. True models are rather expensive, and require a great deal of time and patience to construct, and very accurate work and fine adjustment are necessary if they are actually to make flights.

When the model is at last complete and really flies, an accident may occur, and your beautiful model, which represents weeks of painstaking care, may be smashed to bits in a second. This is very disheartening, whereas a flier, even if broken or smashed—which is unlikely—means merely a small loss of time and money.

The commonest forms of true models which have been worked out on a small scale and which actually fly are the old-style Wright biplane, the Curtiss biplane, the Curtiss hydroaeroplane, the Nieuport, and the Bleriot monoplane. These will be fully described and minute directions for their construction will be given in another chapter.

Before commencing to build a model of any sort, however, great care should be taken to secure accurate and reliable plans and materials. It is very easy to draw out a "scale-plan" of an aeroplane, but it is a far more difficult matter to build one that will actually fly from many of the plans offered for sale at the present time.

## Simple Models

Probably the simplest form of model flier that it is possible to construct is that known as the "skeeter" (Fig. 1). This machine is designed especially for indoor amusement, and consists of a very light, straight stick for the body, or frame,

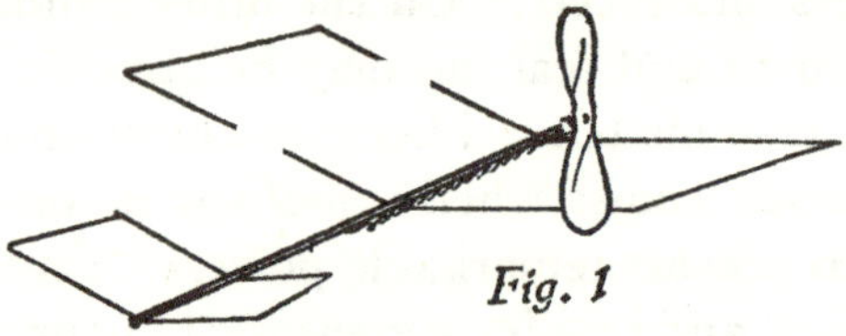

Fig. 1

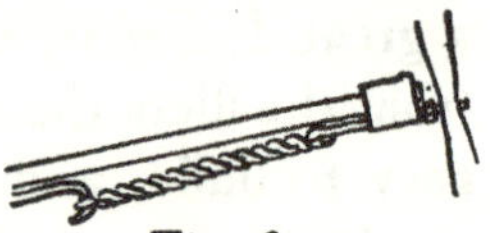

Fig. 2

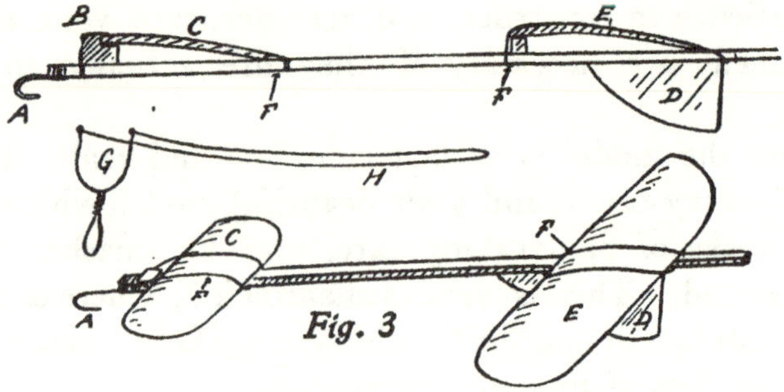


Fig. 3

Fig. 4

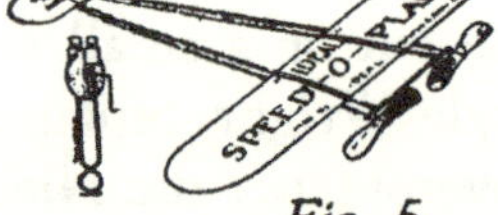


Fig. 5

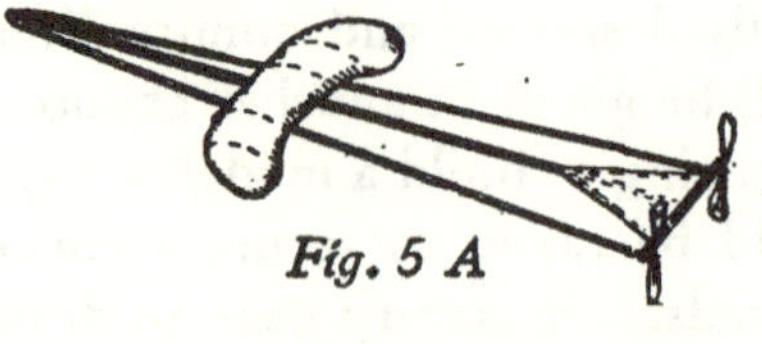

Fig. 5 A

with a forward and rear plane of cardboard or onion-paper. The propeller may be cut from thin sheet celluloid, aluminum, or stiff cardboard, and is turned by a twisted rubber band fastened to a hook on the propeller-shaft and a second hook on the frame (Fig. 2). These tiny affairs are about 8–1/2" long, and spread 6–1/4" across the forward wings. They will fly thirty feet or more, and by bending the wings at various angles and using different-sized planes a great deal of fun may be had, and considerable experience gained as to the particular angle of planes and proportion of sizes which give the best results. These models cost but twenty-five cents ready-made, but any boy can make one in a few minutes at home.

Almost as simple as the skeeter is the "loop-the-loop glider," shown in Fig. 3. For pure amusement this device will prove very satisfactory, for it will fly several hundred feet, and can be made to perform numerous remarkable evolutions, such as "looping the loop," "dipping the dip," etc. This glider consists of a single stick having at its forward end a wire hook (A) and an elevating-block (B) to which is attached the forward wing or plane (C). At the rear end of the stick is a vertical "fin," or keel (D), and above this the main plane (E). Both of these planes, C and E, are attached to the stick by rubber bands (F). Either thin wooden or fiber planes may be used, and even thin aluminum planes often prove very satisfactory. In addition to the glider you must also have a sling of wire (G), to which a long and powerful elastic band is attached (H).

If you wish a straight flight, with the wind, set the front edge of the forward plane on the second notch in the elevating-block and catch the hook (A) on the rubber of the sling. Grasp the rear end of the stick between thumb and forefinger, as shown in Fig. 4, draw back as far as you can, and release the grasp on the stick. If your planes are truly set the little device will sail gracefully upward and away for a hundred feet or more. If a longer flight is desired, you may drive two stakes in the ground several feet apart and use these as the crotch to the sling with longer and stronger rubbers. Two people may also hold the ends of a long, stout rubber, and by one or the other of these methods very long flights of from three hundred to six hundred and fifty feet may be obtained.

Care should always be taken, however, to have the glider start at a spot a little lower than the top of the stakes or sling, and at an ascending angle, in order to have it clear the top of the stakes or the crotch of the sling.

If instead of flying straight the glider flies to one side, set the main plane a *little over to the same side.* A very slight alteration will serve, and it is best to make such changes very gradually until just the proper adjustment is obtained.

By placing the forward or elevating-plane (C) on the upper step of the block (B), and flying the device *against* the wind, the loop-the-loop will be obtained, and the glider will come to earth after a long and graceful glide.

By varying the position of the forward plane up and down on the steps of the block, and by altering the adjustment of the rear plane from side to side, almost an endless number of evolutions may be obtained with this simple device.

These very simple affairs, such as the skeeter and loop-the-loop gliders, are really mere toys, and are useful simply as serving to demonstrate the principle of model machines and as affording an insight into the various effects produced on flight by slight changes in balance, angle of planes, and other details. For real sport or satisfaction in model-flier construction and design you must look for something better, more advanced, and more powerful. Such are the various "long-distance fliers" and "speed-o-planes" (Fig. 5 and Fig. 5 A), the "Cecil Peoli racer," the Pierce, and Mann machines, and a host of other miniature aeroplanes designed for speed, long-distance, and accurate flights, and many of them capable of rising from the ground under their own power.

### Tools and Materials

Many splendid models are for sale ready-made by the various firms dealing in model aeroplanes and supplies, or the machines may be built from accurate plans furnished by these dealers.

As it is far more fun to build your own models and work out the parts from the raw materials yourself, I advise every boy to do this. It is practically impossible to secure the proper grades and qualities of wood, fabric, rubber, and other materials in the ordinary shops, and it is therefore advisable to purchase them from a reliable firm dealing only in such things.

It is possible to build a very good model with only a jack-knife and a brad-awl for tools, and with a piece of pine board, some silk thread, and an old tin can for materials; but such a model built with such crude utensils will never be satisfactory, and is only a makeshift.

Ordinary carpenters' tools are far too coarse and heavy for the fine and accurate work required in building model aeroplanes, and it is far better to provide tools especially suited to the work.

Such tools are not expensive, and save many times their cost in material saved, which would be wasted or injured in attempting to use ordinary tools.

One of the most important tools is a small hand-drill. This is very essential, as it is necessary to drill many small holes in the thin wood of frames and planes, and special drills designed for this purpose may be purchased for twenty-five to thirty-five cents each (Fig. 6).

A pair of small, round-nosed pliers, a pair of small cutting-pliers, a hack-saw or a hand scroll-saw, and a very fine screw-driver are the most important tools in addition to the drills and a sharp knife.

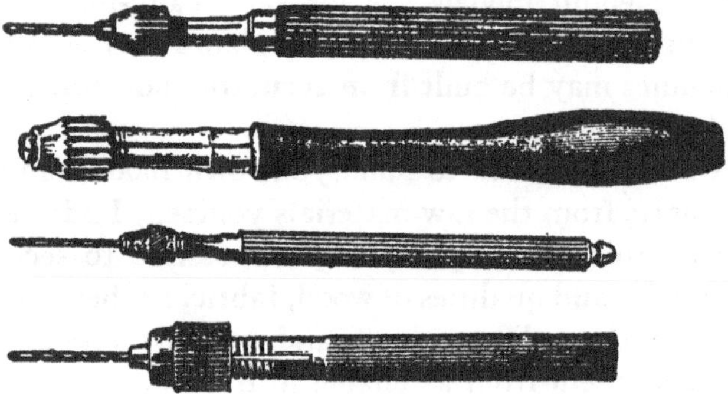

*Fig. 6*

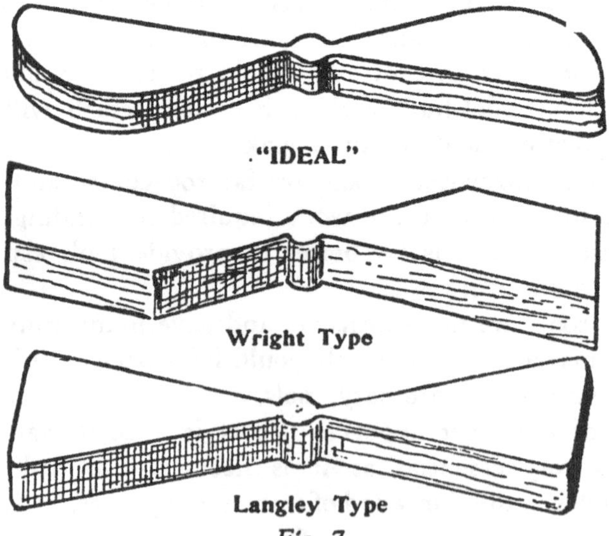


*Fig. 7*

A small iron plane, plenty of assorted sandpaper, strong silk thread, steel and copper wire, sheet brass and aluminum, small screws and brads, and tiny screw-eyes are very useful, and in many cases necessary.

Wire should be flexible, strong, and very fine, gages 28 to 34, and the screws and brads very minute.

In building models lightness and strength combined are of the utmost importance, and common wood with knots or weak spots, ordinary cardboard or paper, and common elastic bands will *not* serve if good results are to be obtained.

Among the fabrics used are "silk fabric," "bamboo-paper," and "fiberloid," all excellent for certain purposes. The silk and bamboo fabrics should be coated with bamboo varnish or fabric solution; but fiberloid is a celluloid-like transparent material and requires no treatment.

Other fabrics can also be obtained which are already treated and require no varnishing or other preparation.

Among these is "Zephyr Skin," a specially prepared material which is lighter than tissue-paper and tougher than silk, and which has an exceedingly smooth surface, thus reducing skin friction to the minimum. The size of the skins is about 6″ x 22″, but two or more may be readily glued together to obtain any desired size.

To those unfamiliar with model aeroplane work and the advances made in model-aeroplane construction a great surprise is in store when a visit to a dealer in model supplies is made.

In addition to selected materials, such as woods, fibers, bamboo, rattan, metal, etc., there is a great variety of dainty carved propellers, metal propellers, and fiber propellers for machines of every sort and size.

Shafts, bearings, metal braces, and even ball-bearings, are constructed solely for models, and an infinite variety of tiny rubber-tired wheels, axles, turnbuckles, gears, sprockets, and other parts are to be had ready-made, and as carefully finished as their larger counterparts used in real aeroplanes.

In many cases better results may be gained by purchasing all such accessories ready-made than by attempting to fashion them by hand, for they are wonderfully cheap, and unless you are well provided with a fine assortment of metal-working tools it is a difficult matter to produce accurate metal fittings for models.

Propeller hangers, bearings, braces, and similar fittings may be obtained to suit almost any model; but, as most of these are adapted to special designs, it is better, in experimental work, to make your own metal parts.

The only materials which should be kept on hand are various sizes of spruce sticks, T- and I-beam section wood, round and half-round rattan or reed, and split bamboo. As you can never tell just what size or form of wood, rattan, or bamboo you will require, it is best to keep a good assortment on hand, for these materials are cheap, and a great deal of time and trouble will be saved if you have just the thing you need and are not obliged to run out or send for a piece of wood you happen to want.

The spruce in various forms costs but one or two cents a foot, and the rattan and bamboo about the same, so you can have a splendid assortment for less than a dollar.

If you wish to make your own propellers you should have several blanks on hand, as these are very cheap, costing from ten to twenty-five cents each, and are far better than you can make yourself. It is an easy matter to whittle a single propeller from a blank; but it is very hard to whittle out two exactly alike where twin propellers are used. The blanks are made of various suitable woods, and are roughly shaped, as shown in Fig. 7, which illustrates some types of "Ideal" propeller blanks.

Equipped with the above-named tools and materials, the boy builder need not be afraid to tackle any sort of model, for with industry, care, and accurate work splendid models can be built even if you have never practised carpentry or wood-working.

Of course, more attractive and handsomer models can be produced by using ready-made parts and merely assembling them, but the satisfaction obtained by building an aeroplane model entirely by yourself is far greater than you will obtain by putting together parts purchased of a dealer.

A model aeroplane is much like a model boat or yacht; the boy with plenty of spending-money can produce results very elaborate and completely equipped, but the boy who makes every part himself feels an interest in his finished product, and knows its whims and peculiarities, and consequently gets better practical results from his model than if he had spent more money and less hard work and painstaking care in its construction.

## Chapter V

### HOW TO BUILD RACING MODELS

THE model flier which holds the world's record for distance or duration is always of the greatest interest to model-flier enthusiasts. Like real aeroplanes, however, the records are so often broken and new ones established that it is very difficult to keep track of just which one is really the record-holder.

Moreover, official and unofficial records may differ, and whereas an unofficial record may be correct, yet unless it is official it will not stand as a record among the clubs and model enthusiasts.

A record flight to be official must be made in actual competition in a duly announced contest, and under accepted rules and conditions of model contests.

Many of the longest recorded flights are unofficial, among them the flight of the Mann monoplane, which is alleged to have covered 4,200 feet with a duration of 100 seconds. Cecil Peoli has also made various flights of 2,500 feet and more unofficially, and the Pierce record of 1,814 feet 6 inches was also unofficial.

For a long time the Cecil Peoli racer was the holder of all American records, with a flight of 1,691 feet 6 inches, and a duration of 48–5/8 seconds.

The official records at the present time are as follows:

WORLD'S MODEL FLYING RECORDS

| | | | |
|---|---|---|---|
| Hand-launched | Distance | Armour Selley | 2,653 feet |
| | Duration | Armour Selley | 158–4/5 sec. |
| Off ground | Distance | Armour Selley | 1,408 feet |
| | Duration | Curtis Myers | 73 sec. |
| Hydro off water | Duration | G. H. Cavanaugh | 6–2/5 sec. |
| | | | England. |
| Single tractor screws | Distance | H. R. Weston | 252 feet |
| | Duration | F. W. Jannaway | 22 sec. |

AMERICAN MODEL FLYING RECORDS

| | | | |
|---|---|---|---|
| Hand-launched | Distance | Armour Selley | 2,653 feet |
| | Duration | Armour Selley | 158–4/5 sec. |
| Off ground | Distance | Armour Selley | 1,408 feet |
| | Duration | Curtis Myers | 73 sec. |
| Hydro off water | Duration | G. H. Cavanaugh | 60–2/5 sec. |

BRITISH MODEL FLYING RECORDS

| | | | |
|---|---|---|---|
| Hand-launched | Distance | A. E. Woollard | 1,431 feet |
| | Duration | A. F. Houlberg | 89 sec. |
| Off ground | Distance | G. Roulands | 696 feet |
| | Duration | A. F. Houlberg | 51 sec. |
| Hydro off water | Duration | G. P. Bragg Smith | 25 sec. |
| Single tractor screw | Distance | F. G. Hindsley | 519 feet |
| Hand-launched | Duration | J. E. Louch | 44 sec. |
| Single tractor screw... From ground | Duration | J. E. Louch | 40 sec. |

(All British records are quoted from *Flight*.)

The Mann monoplane, whether its record is official or not, is a very efficient machine designed by an English schoolboy. Its total weight ready for flight is but four ounces; it flies at about eighteen miles an hour in favorable air, and it is very simple and an easily constructed model flier.

The machine consists of a triangular body or fuselage (Fig. 1, A) 34 inches long and tapering from 17 inches wide at the rear to a point in front. The wings or planes are placed at points two inches from the forward end and 12 inches from the rear end, and the whole machine is driven by twin propellers of eight-inch diameter.

The triangular frame consists of two pieces of selected silver spruce 1/4″ square and 34 inches long, with a cross-brace at the rear end of 1/4″ x 1/8″ spruce seven inches long.

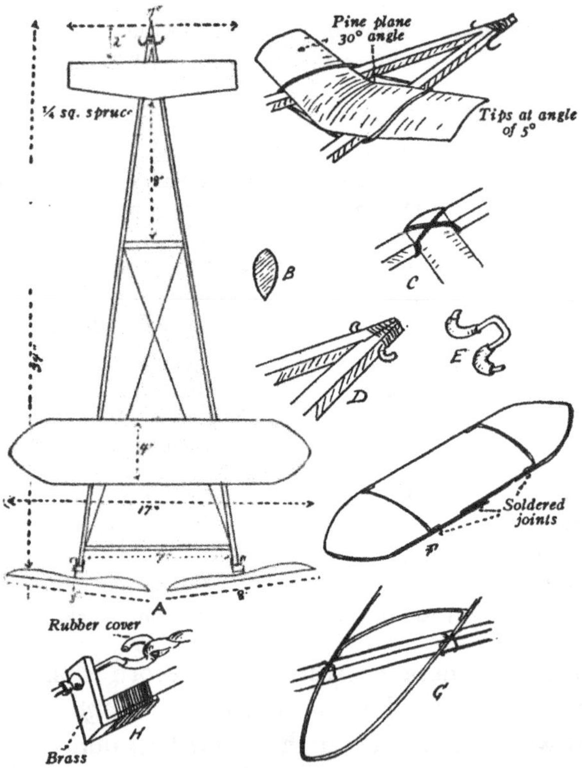


*Fig. 1*

This rear cross-brace is of stream-line section (Fig. 1, B), to reduce air resistance, and is lashed to the main frame by silk thread glued or shellaced (Fig. 1, C). A second cross-piece is also lashed and glued on the top of the frame twelve inches from the pointed forward end of the frame, and in shape and size is similar to the rear brace. In gluing these cross-braces in place only the finest white glue should be used, and the thread lashings should be wound evenly and tightly, and then coated with a layer of glue and shellac. The two forward ends of the main beams should be cut off at an angle, as shown in Fig. 1, D, and glued together. A curved W-shaped piece of stout wire four inches long is set over the end in a slight notch, and is lashed and glued as shown in Fig. 1, E, thus holding the two ends of the beams together and serving as a hook to which the rubber strands are fastened. The two curved sides of these hooks are covered with a piece of rubber tubing. From the junction of the forward cross-brace with the side-beams wire guys extend diagonally to the junction of the rear ends of the main beams with the rear brace. These guys are of very fine piano-wire set taut with small turnbuckles, which may be purchased at any dealer's in model-aeroplane supplies (Fig. 1, A). The main or rear plane is elliptical in form, with a total span of 17 inches and a chord of four inches (Fig. 1, F). The frame of this plane is made of 18-gage piano-wire bent to the proper shape and soldered together. In making this frame it will be much easier if a wooden form or pattern is first cut out and the wire bent around it. By using 1/2" board for the pattern the wire may be held in position while soldering by tacking small staples over it. The plane has two cross-ribs, as shown in the figure. These are 18-gage wire also, and they are bent or curved upward to give the plane an arch or camber of 1/2". The frame of the plane should be fastened to the main frame by lashings of silk and shellac, as illustrated in Fig. 1, G. The most difficult part of the plane construction is in covering it with fine waterproof silk, which is cut to shape and then laced taut

Photo by Frank Wiehn

MODEL HYDROAEROPLANE RISING FROM THE WATER, OAKWOOD HEIGHTS, STATEN ISLAND

Photo by M. Palmer

HAND-LAUNCHING A FLYING MODEL, OAKWOOD HEIGHTS, STATEN ISLAND with strong silk thread. You will probably have to try several times before you succeed in getting the silk just the proper size and in adjusting it in such a way that it will fit the frame snugly and smoothly and yet not wrinkle or tear. It is next to impossible to describe this adjustment, but a few trials will help wonderfully well, and by using some cheaper and commoner material for your experiments you will soon learn the knack of getting the covering on smoothly.

The forward or elevating-plane is 7" long x 1-3/4" wide, and is made from spruce 1/30" in thickness. The forward edge of this plane is straight when seen from above, with the rear edge tapering from the center to the square ends, which are one inch wide (Fig. 1, A). The two sides of the elevating-plane tip up from the center at a dihedral angle of 30 degrees, while the rear tips bend down at an angle of about five degrees, and upon the curve of these tips depends the elevation of the machine. The plane is easily given the desired curve by steaming and bending, and by making a form of wood and fastening the steamed plane on this until dry and cold very accurate results may be obtained. When thoroughly dry the wood should be carefully sandpapered with the finest sandpaper until perfectly smooth. It is fastened to the frame by rubber bands, which are slipped over the frame, and the free loops passed over the projecting ends of the plane. As this thin spruce elevator is very fragile, it should be placed in position only when in use, and several extra elevators should be provided so that if one breaks another may be substituted. After sandpapering, the plane should be given two coats of fine shellac. The spot where this plane is to be placed (two inches from the front of the frame) should be marked in pencil on the main frame. The propellers are made of birch or other hard wood 1/20" in thickness, and are eight inches in diameter, with a pitch of 24 inches. Instead of being cut from a block of wood as usual, they are bent from a thin strip by steaming it.

As it is very difficult to secure an accurate pitch in this way, or to get both propellers exactly alike—which is essential to good flights—it is better to buy ready-made propellers such as the "Ideal," which may be obtained at a cost of from 75 cents to $1.50 a pair, and are absolutely true, and are sold in pairs. These may be purchased of plain wood, laminated wood, fiber, metal, or aluminum, and will prove far more efficient than anything you can make. The bearings for the propeller-shafts are merely L-shaped pieces of brass with one side fastened to a side of the frame and the other arm bored with a small hole to receive the propeller-shaft, as shown in Fig. 1, H. Between the brass piece and the hub of the propeller are two steel collars, or washers, to reduce friction. One of these is stationary, while the other revolves, and they should be well lubricated with vaseline.

A great improvement over this rather crude construction can be made by substituting factory-made propeller-shafts with turned or ball bearings. Turned bearing-shafts cost 15 cents each, while ball-bearing shafts cost 50 cents each, and for ordinary work the turned bearings answer every purpose (Fig. 2). The brass shaft-hangers should be fastened in place by shellac and tiny brads, and should also be lashed firmly to the frame by silk thread coated with shellac. The end of the shaft is bent into a hook and covered with a piece of rubber tube to protect the strands of rubber used as a motive power. Each motor consists of six strands of 1/4" flat elastic looped over the forward wire hook and the hook on the shaft, and they should be given about 1,000 turns to produce a propeller-speed of 750 revolutions per minute. As it is a tedious job to wind up propellers by hand, it is advisable to purchase or make a winder. The Ideal patented friction-winder (Fig. 3) is a very good form, as this insures even winding of both propellers. Another good form of winder is shown in Fig. 4. This can easily be made from an old Dover egg-beater by cutting off the beaters and replacing them with wire hooks as shown. When winding the rubber strands, for this or any other model, with a winder

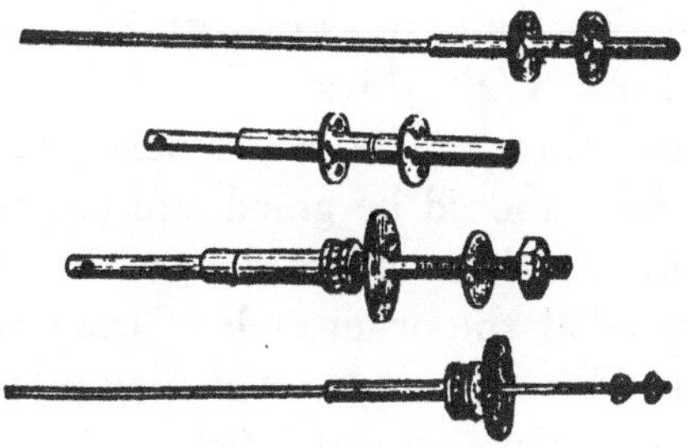

*Fig. 2*

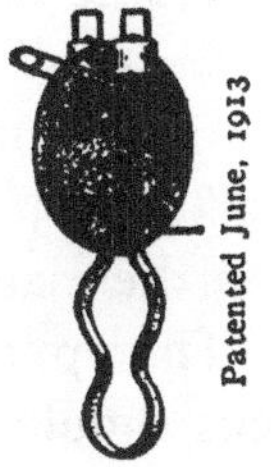


*Fig. 3*

*Fig. 4*

the rubber strands should be stretched to twice their length. Owing to the fact that the sizes of rubber strands did not always agree and confusion often occurred in ordering, the dealers have now adopted the plan of listing and selling rubber by number. Number 16 was formerly known as 1/16″ rubber and Number 14 as 3/32″.

If the machine is to be used entirely for racing or competition flights light weight is most desirable, and no further attachments should be added; but if the machine is to be used for pleasure and sport, a landing-gear or skids of some sort should be added. These may be made of light strips of reed or split bamboo 3/32″ square and lashed to the lower side of frame. One of these should be placed at the forward end, and one on each main brace near the rear.

## How to Build a Peoli Racer

A splendidly efficient and easily built machine is the Cecil Peoli racer, invented and produced by Cecil Peoli, of New York City, who is the youngest professional aviator, and who has produced several novel and efficient gliders and models, as well as full-sized aeroplanes. This machine, like the Mann monoplane already described, is of triangular form, with a frame made of two long sticks of spruce, each 34″ long and 3/16″ wide and 1/4″ thick. To the rear end of each stick, on the outer side, are bearing-blocks 3/4″ x 1/4″ x 1/2″ (Fig. 5, *aa*), which should be glued and lashed in position on the main beams, on their widest side, as shown in the cut. The opposite sides at the other ends of the two fuselage sticks should be cut off at an angle until they meet evenly with the rear ends five inches apart (Fig. 5, Plan). Across this triangle are three bamboo braces (Fig. 5, c), each 1/16″ thick and 1/8″ wide, lashed in place with silk thread as usual. Through the two side-beams, where they join in front, a 1/16″ hole should be bored and a stiff wire placed through this hole and bent in a hook at each side, as shown in the plan at *dd*. The two ends of the main frames should then be lashed and glued as usual. Thin bamboo strips 3/32″ wide should next be bent into the form shown in Fig. 5, C1, C2, and lashed to the lower sides of the main beams—one in front and two at the rear.

It is very easy to bend bamboo, rattan, or thin wood by steaming or heating over a spirit-lamp, and you should never attempt to bend such materials either dry or cold. In placing the rear skids (C2) in place care should be taken that they slant slightly outward, as shown in the front view in the figure. When the correct slant is obtained the cross-piece of bamboo (*f*) is lashed in place and the V-shaped

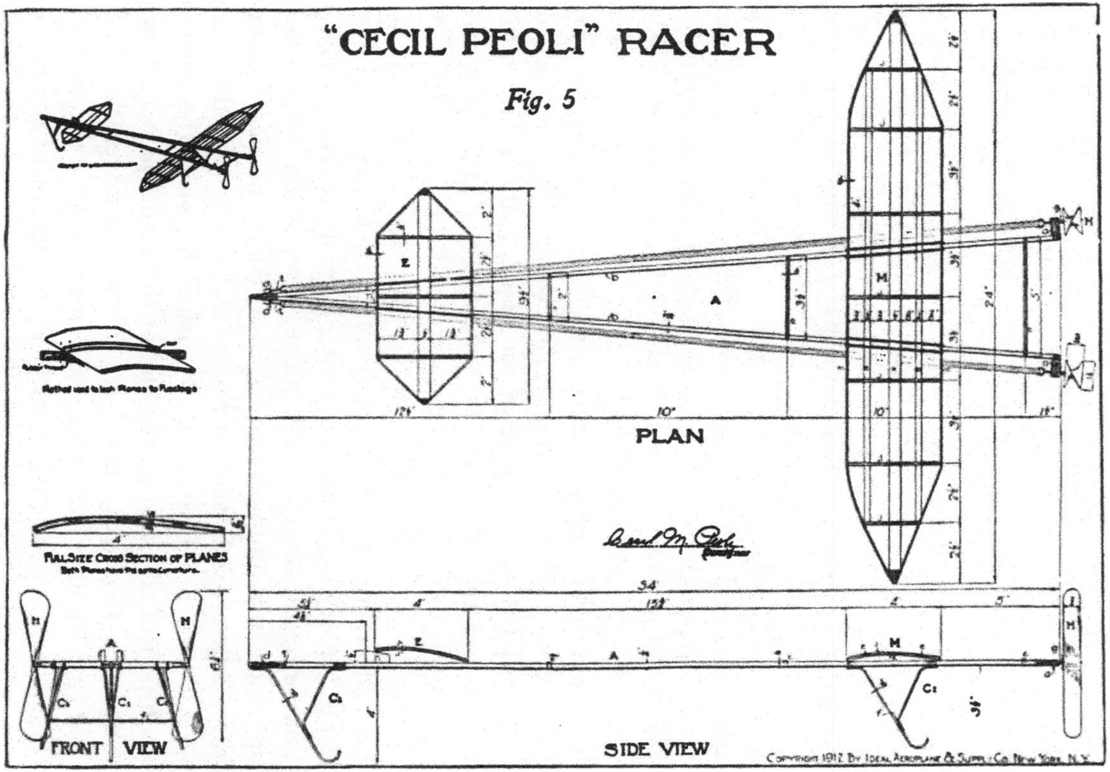
"CECIL PEOLI" RACER
Fig. 5
Method used to lash Planes to Fuselage
FULL SIZE CROSS SECTION OF PLANES
Both Planes have the same Curvature.
PLAN
SIDE VIEW
FRONT VIEW
Cecil M. Peoli
COPYRIGHT 1912 BY IDEAL AEROPLANE & SUPPLY CO. NEW YORK, N.Y.

space on the skids is covered with bamboo-paper and given a coat of bamboo-varnish.

The main plane should now be constructed, and in this work great care should be taken, as upon this depends the success of your model.

For the ribs there are fourteen pieces of 1/32″ bamboo, which should be bent to the curve shown in the cross-section of planes, and these should be lashed, two at a time (upper and lower), to the main spar (*k*), which is 24″ long, 1/2″ wide, and 1/16″ thick, at their centers. All the ribs should be bound in place accurately at right angles to the main spar and spaced as follows: Outer ribs on each side, 2–1/2″ from end of main spar; second ribs on each side, 2–1/2″ from first; third ribs, 3–1/2″ from second; center ribs, 3–1/2″ from third.

When this is accomplished slip the two auxiliary spars (*ee*), each 1/4″ wide and 1/16″ thick, between the ribs from one end to the other of the plane, and slide them into position so that they will be spaced 1–3/4″ from the front edge, and 3/4″ from the main spar, and the other 5/8″ from the main spar and 7/8″ from the rear edge. Next glue and lash all the joints of ribs and spars firmly in place, and while these are drying build the front plane or elevator. This is built exactly like the main plane, except that no auxiliary spars are used and only three sets of ribs are needed. Around the edges of each plane strips of bamboo 1/16″ thick and 1/16″ wide are bent, and are lashed between the ends of each pair of ribs and at the end of the main spar, as shown. Cut some bamboo-paper large enough to cover the plane, and turn up over the edges and cement or glue this in place. Then cut a second piece just large enough to fit the plane's top, and glue this firmly over the edges of the bottom piece. When dry the paper should be given a coat of bamboo-varnish.

When dry the planes may be attached to the main fuselage by rubber thread, as shown; but before placing the elevator in position two small blocks, each 3/16″ thick and 1/2″ wide at the widest end and 3/32″ wide at the other and 1/2″ long, should be attached to each main-frame beam 4-7/8″ from the forward end. The front edge of the elevator should then be adjusted to rest on the sloping side of these blocks at 5-1/4″ from the apex of the fuselage. The propellers are 6-1/4″ in diameter, 1-1/4″ wide, and 3/4″ thick, and may be cut from blocks by hand, or may be bought ready-made, which is advisable if the best results are to be expected. The shafts pass through 1/16″ holes in the bearing-blocks. Small hooks of steel wire should be made, as shown at *t*, and 18 feet 8 inches of 3/32″ x 3/32″ rubber should be wound in a skein of seven strands from each of these hooks to its respective propeller-shaft hook. Your racer is now ready to fly, but you should wait for a calm day before trying it, as at first there very likely will be numerous small adjustments to make before a straight and reliable flight is accomplished.

By the aid of a propeller-winder, such as already described, or by winding each propeller the same number of times by hand (for about 600 to 800 turns), the rubbers may be twisted taut. Hold both propellers in the right hand and raise the model above the head with the left hand. Let go with the left hand, and at the same time give a slight push with the right.

If the machine flies upward or "climbs" too rapidly remove the elevator and cut off about 1/16″ from the upper surface of the elevating-block. Continue this until the proper angle of ascent is attained.

If the model dives, whittle the elevating-blocks down until you get more elevation. If it shows an inclination to turn in circles, the main plane is out of center or warped, and it must be bent into shape or adjusted so that a steady flight is attained. For delicate front and rear stability adjustment move the main plane a little (half an inch or so) forward or back. If pushed forward the machine will climb more, if pushed back it flies on a more level course. It is best to mark the position when a satisfactory flight is secured, in order to avoid unnecessary adjustments every time.

In speaking of fastening various parts I have mentioned "glue," but, while good white glue is an excellent material, it is subject to weather changes, and a waterproof cement such as Ambroid is preferable. Bamboo-varnish and bamboo-paper are materials especially made for model-aeroplane use, and may be purchased of any dealer in model supplies.

### The Pierce Model

Another very fast, powerful, long-distance model, which for a long time held the American record for long flights, is the Pierce monoplane.

The unofficial record of this machine is 1,814–1/2 feet, with a duration of 72 seconds. It is a simple, easy model to build, and every boy who takes up model-building should certainly build a Pierce.

The more styles and types of models you have, the more likely you are to learn just why and how one model is superior to another, and by combining the best points of several you may perhaps invent an entirely new and remarkable flier.

The Pierce is a triangular machine composed of two sticks of straight-grained spruce each 34″ long and tapered from 3/16″ at the middle to 1/8″ at the ends.

The two forward ends are shaved down and glued and lashed together in the same way as in the Mann or the Peoli model, and the apex of the frame carries a steel-wire hook as shown in Fig. 6.

This hook is lashed in place by silk thread, and two small S-hooks are used to connect the rubber strands, as illustrated in Fig. 7.

The rear cross-piece is of spruce 1/32″ square, shaved to stream-line cross-section, and the twin propellers are attached to this.

The planes are built up of bamboo strips bent around a light spruce frame, as shown in Fig. 8, and are covered with

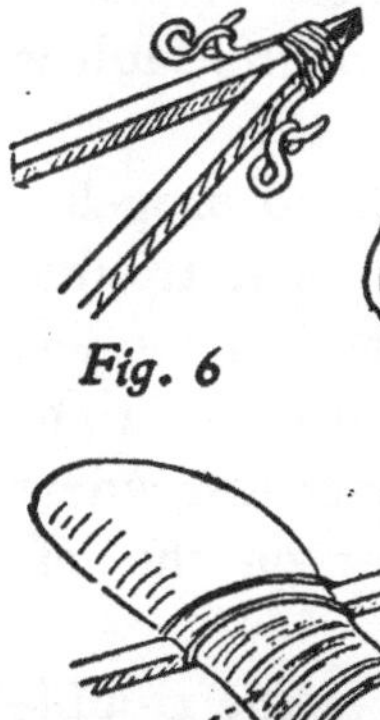

*Fig. 6*

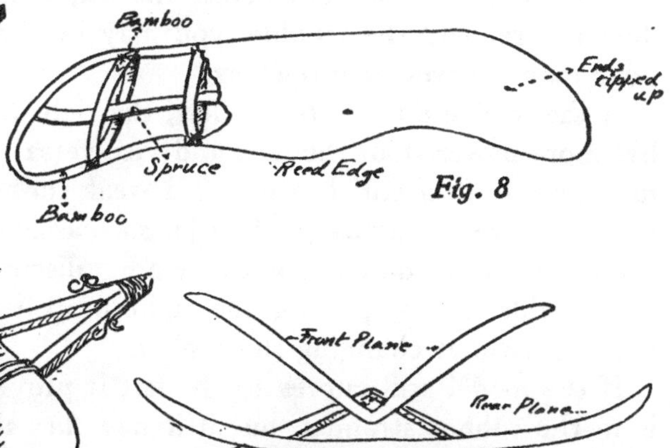


*Fig. 8*

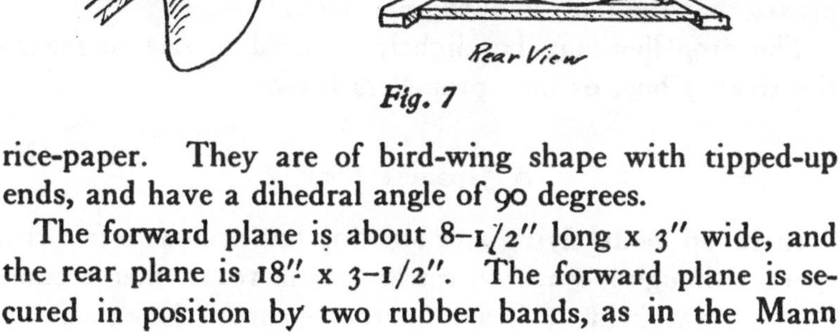


*Fig. 7*

rice-paper. They are of bird-wing shape with tipped-up ends, and have a dihedral angle of 90 degrees.

The forward plane is about 8–1/2″ long x 3″ wide, and the rear plane is 18″ x 3–1/2″. The forward plane is secured in position by two rubber bands, as in the Mann model, and should be about three inches from the extreme front end of the frame.

The rear plane is 12 inches from the rear end of the frame, and eight-inch twin propellers are used, each driven by eight rubber strands.

In this as in all other twin-propeller machines great care should be taken to have the two propellers of exactly the same diameter, pitch, and speed, and also to have them revolve in opposite directions.

With home-made propellers it is difficult to accomplish this, but with ready-made propellers little difficulty will be encountered.

If your model flies crookedly and adjustment of planes does not remedy the trouble, you may be sure the fault is your motor-power or propellers.

If the machine turns to the left, the right-hand propeller has more power than the left, and vice-versa. The trouble may be due to a stiff bearing or a weak rubber on the left propeller, or to a higher pitch or larger diameter in the right propeller. If in doubt, you can readily discover just where the trouble lies by placing the right propeller on the left bearing without changing the rubbers.

If the model still curves to the left it proves the trouble is in the rubber strands; but if it now flies to the right, it proves that the fault is in the propeller.

The propeller may be slightly reduced in size to match the weaker one, or new propellers fitted.

### A Japanese Flier

In Japan model flying and building has become a favorite sport, although it was first introduced in 1911, when a competition was held at Osaka in the beautiful Nakanoshina

Park. At this meet a model built by K. Nakagawa won the prize by flying 105 yards in 12 seconds.

The Japanese boys and young men at once became enthusiastic over models, and there are now over a dozen clubs, and numerous meets are held yearly.

The Japanese are noted for delicate and accurate work, and in the construction of model aeroplanes they discovered a field in which their skill in handiwork proved most valuable.

Some of the Japanese models are very beautifully and carefully made, and while, as a rule, American models are followed, yet many are very original, and are well worthy of imitation by American designers.

The *Angel IX.* of Mr. Nakagawa is a typical example of prize-winning Japanese models.

This is a monoplane of triangular frame, and, while the original is composed of native Japanese woods, spruce and other American products will serve just as well.

The main beams are 3/16" x 1/4", each 36 inches long, and are connected by cross-beams (Fig. 9, A, B). The main beams, instead of being square or rounded, are of I-beam section (Fig. 10), which insures great strength with extreme lightness, and affords an excellent means of attaching the cross-beams and other parts in a neat and workmanlike manner.

The two cross-bars mentioned (Fig. 9, A, B) are spaced 12 inches from the front and five inches from the rear, and are of stream-line section with the ends fitted snugly into the grooves in the main beams (Fig. 11). A thin strip of sheet aluminum is then bent around the main beam and over the cross-braces, and is held in position by glue and silk-thread lashing. These strips of metal are but 1" x 1/8" in size, and the weight is so slight that it is of no consequence,

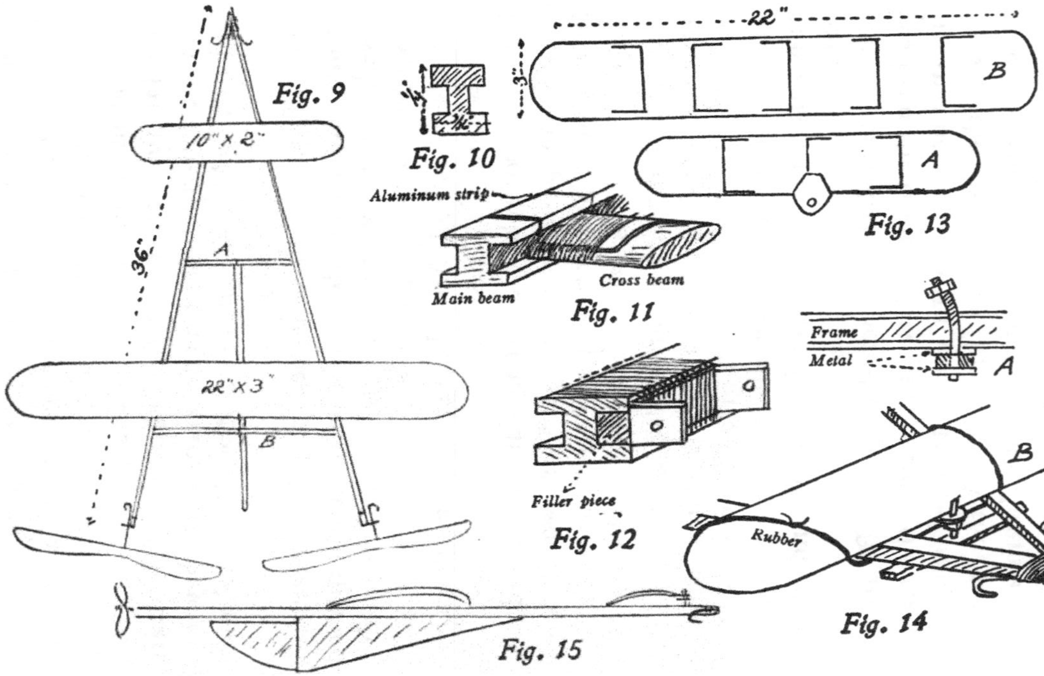


A JAPANESE FLIER

whereas the superior strength and rigidity gained by this method of attachment is of great value.

The lightness obtained by using I-beam section main beams allows the use of metal in many important details without an increase over the normal weight of the ordinary model.

At the extreme rear ends of the frame-beams are the bearings for the propeller-shafts. These are made of 1/32" steel bent as shown in Fig. 12, and attached to the outer sides of the main beams by silk winding and glue.

A thin wooden block should be glued into the groove on the beam beneath the plate, as indicated, before winding on the bearings.

The propeller-shafts are of No. 15 steel wire, and they are fastened to the propellers by a fine wire pin passing through a hole drilled through the propeller and the wire shaft.

It is a mighty difficult and delicate job to drill a hole through a No. 15 steel wire, and few American boys could accomplish it, and I should therefore advise the use of an ordinary shaft. If you wish fully to appreciate the exquisite accuracy of the Japanese construction you may try your hand at thus drilling the wire; but I assure you that you will find it most difficult.

The propellers are 10 inches in diameter, with 15-inch pitch reduced to 10-inch pitch near the hubs. The original model carries hand-made wooden propellers carefully whittled and scraped to perfect balance, and coated with five coats of varnish and highly polished. Ordinary ready-made propellers will serve every purpose, however.

In the original model a round glass bead was placed between each propeller-hub and its bearing; and, while this serves very well when the simple shaft is used, yet regular ball-bearing shafts or plain bearings will answer as well or even better.

The apex of the frame is provided with a hooked wire, as usual, for attaching the rubber strands, 1/16″ square, with twelve on each side.

The planes are of light Japanese silk coated with shellac, and are stretched over piano-wire frames. The large plane is 22″ x 3″, with five ribs of wire bent as shown in Fig. 13. The bent ends of each rib are then wound to the edge-frame by fine copper wire and are soldered firmly in place.

The forward or elevating plane is of the same construction, 10″ x 2″, with three ribs, and on its forward edge is a thin metal plate with a 1/16″ hole, as shown in Fig. 13, A.

This plate slips over a 1/16″ bolt attached to a short cross-brace on the lower side of the frame, and rests upon a nut on the bolt. The plane is attached to the frame by a rubber band, and by merely screwing the nut up or down any desired elevation is easily obtained (Fig. 14).

The bolt is fastened to the block by soldering to thin pieces of metal on the upper and lower sides, as shown in Fig. 14, A, and the block is attached to the frame by glue and silk thread as usual.

Across the top of each plane a stout silk thread is stretched from tip to tip, thus giving a dihedral angle and trussing the planes also.

Beneath the frame a keel or fin is placed extending from the forward cross-brace to the rear extremity of the main beams (Fig. 15).

This keel serves to stiffen the model, protect the propellers when landing, and aids in producing straight flights, and adds to lateral stability.

This keel is attached by elastic bands, and the rear end

acts as a rudder, which may be slightly moved from side to side to correct any tendency to fly to right or left.

### The Lauder Duration Model

This efficient and simple flier won the duration contest at Murray Hill, New Jersey, October 20, 1912, and, while the record has since been broken, yet its performance at that time was remarkable.

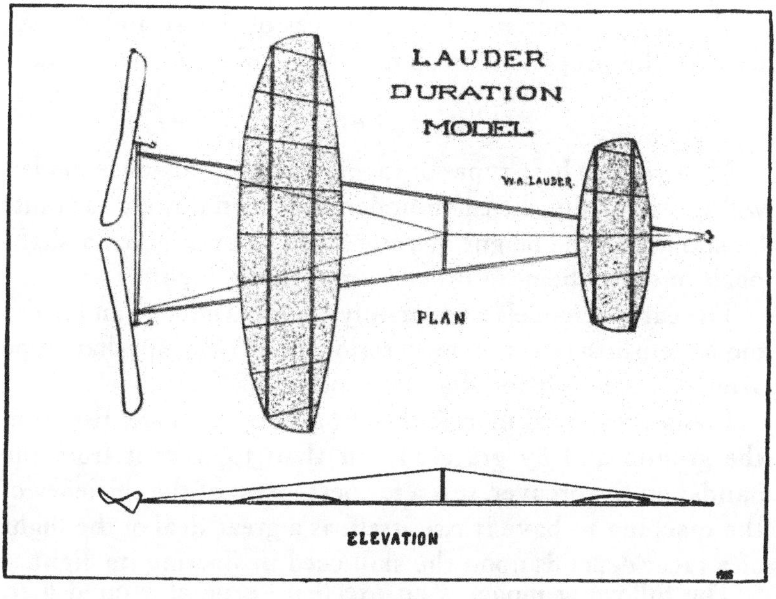


*Fig. 16*

Its record flight was 143 seconds; but eight times, in eight consecutive flights, this model flew over 130 seconds.

The construction of this model is so simple that scarcely any detailed directions are necessary, and its appearance is fully shown in Fig. 16.

The fuselage is 40 inches long, and is of silver spruce

1/4" x 5/16", tapered toward the ends. The main beams are of oval section, carefully smoothed and varnished. The front .plane is 13" x 5", and the main or rear plane 27" x 7", built up of bamboo and covered with Japanese silk paper.

The propellers have a blade surface of 14–1/2 square inches each, and a pitch of 34 inches. They are 12 inches in diameter and of true screw shape.

The machine is guyed with No. 34 steel wire, and weighs, complete, 4–5/8 ounces.

The motor consists of two ounces of flat-strand rubber, driving the propellers at about 500 revolutions per minute.

### Ground-Fliers

Many of the best types of racing and long-distance models will not rise from the earth under their own power, but must be started at the height of your head, or even given a slight push to start them, in order to make long flights.

The earliest models were mainly "ground-fliers," and many model enthusiasts are now seriously taking up that type after discarding it for the more speedy racing models.

It is a great deal more interesting to see a model rise from the ground and fly gracefully off than to start it from the hands; and, moreover, it is a far better test of the efficiency of the machine to have it rise itself, as a great deal of the flight of a racer depends upon the skill used in starting its flight.

The following model is an excellent type of ground-flier, but almost any well-designed model will rise from the ground if given larger planes, skids to rest upon, and the right elevation to the front plane.

For building the ground-flier you will require the following materials:

Two pieces of spruce, each 35" long x 1/4" x 5/16".
Two 6" propeller-blanks or two 6" ready-made propellers.

Thirty-six feet 3/32" square rubber.
Four feet of split bamboo.
One foot 1/16" steel wire.
Two and one-half feet thin spruce 1/16" thick.
Two propeller-bearings and shafts.
One can Ambroid varnish.

The above will cost from $1.50 to $3, according to the quality of shafts, propellers, and bearings used.

To build the frame cut off the side of each long stick at one end and lash and glue them together as shown in Fig. 17.

About 1/2" from the other ends lash and glue a brace of split bamboo to the upper side of the sticks, and fasten another in the same way half-way between the rear brace and the apex of the frame (Fig. 18, A, B).

Cut two small wooden blocks in the form of steps (Fig. 17, C), and glue and lash these to the top of the frame sticks three inches from the forward end.

Through the two sticks, where they are joined in front, bore a 1/16" hole, and through this place a steel wire bent in the form of a hook on each side (Fig. 17, H).

The bearings are glued and lashed to the outer side of the rear of the main beams, and may be home-made of thin metal bent and bored as described for the Japanese flier, or they may be purchased ready-made for a few cents.

The machine carries six-inch propellers; and, while these may be whittled from blanks at home, yet better results may be obtained by purchasing them ready-made.

Beneath the model are skids of split bamboo constructed as follows:

Heat a piece of bamboo 5-1/2" long over a spirit-lamp or flame, and bend it as shown in Fig. 19. Then split this in half, and you will have two skids exactly alike.

The braces to these (Fig. 19, A) are made in the same way, using pieces 4-1/2" long.

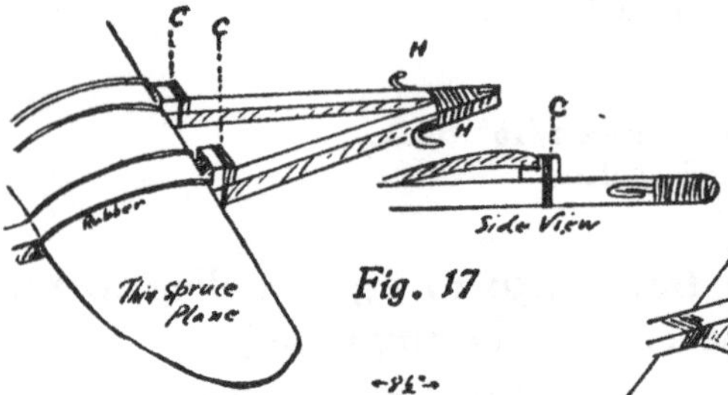


Fig. 17

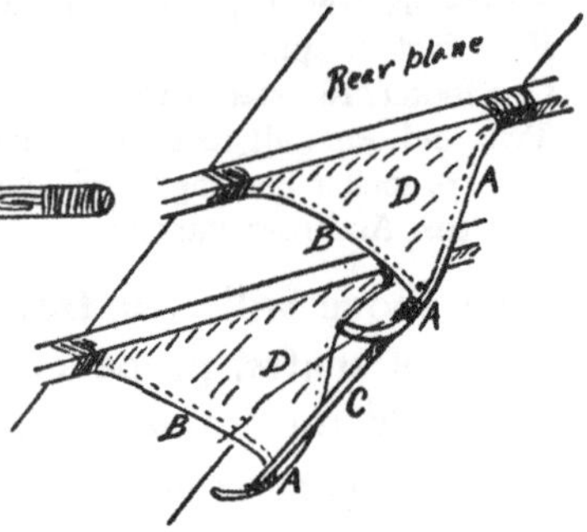


Fig. 19

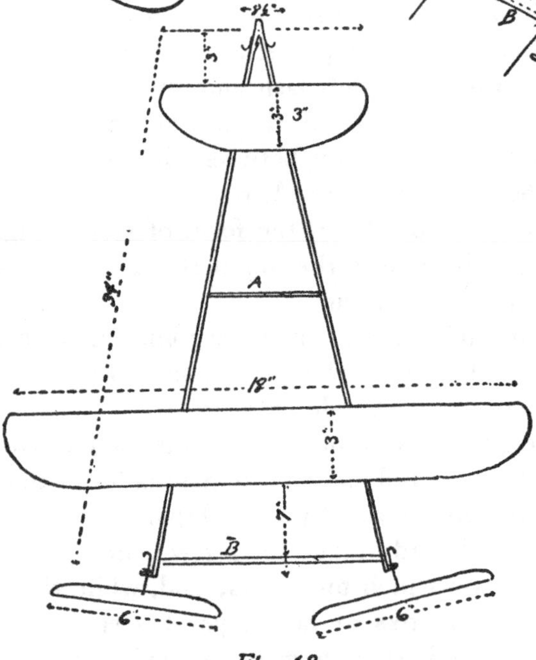


Fig. 18

The braces are lashed and glued to the skids, and both skids and braces are lashed to the frame, with the upper end of the braces four inches from the rear end of the beams, and the front upper ends of the skids seven inches from the rear, or, in other words, 2–1/2″ from the ends of the braces (Fig. 19, A, B).

From one skid to the other a thin strip of bamboo is lashed at the spot where the braces join the skids, in order to steady them (Fig. 19, C).

The front skids are made like the rear skids, but the forward skid is 7–1/2″ long, and the braces 6–1/2″, in order to give the machine a slight upward tilt, or angle of incidence, and thus allow it to rise.

After the skids are attached they are covered with bamboo-paper and varnished, and then act as keels or fins, increasing stability and producing straighter flights (Fig. 19, D).

The planes are made from the thin 1/16″ spruce, the main plane being 18″ x 3″ (Fig. 18), and the front or elevating plane 8–1/2″ x 3″. Both planes are bent or curved slightly by heating the wood and bending it into shape.

The planes are lashed to the frame by rubber strands, with the forward edge of the elevator resting on the two wooden steps (Fig. 17).

Each motor should have six strands of the rubber, which should be attached to the wire hook at the forward end by small S-hooks so that the rubber may be easily slipped off for winding with a winding-machine such as already described.

In winding these rubbers not over 250 turns should be given, or the strands may break.

In winding rubber strands for this or any other model the rubber should be stretched out while winding, as otherwise it will kink, and one person must hold the model while the other winds.

After it is wound the rubber is hooked over the front wire hook and the machine placed on smooth ground while the propellers are held by the hand.

As soon as the propellers are released the machine will slide along the ground a short distance and then rise gracefully in flight like a real aeroplane.

If it fails to rise, or uses up most of its power before rising, the front plane should be elevated more by raising it a step on the blocks, or the rear plane should be moved forward slightly.

Instead of the fragile wooden planes you may use planes made of a wire or bamboo framework, with bamboo or silk-fabric covering, such as those described for the Mann, Pierce, Peoli, or Japanese models. The wooden planes are merely easier to build; and, as the machine is not intended for racing, they answer very well.

Moreover, for experiments you may use wooden planes of various sizes and curves, and when you decide on the very best proportions you may easily duplicate them in a frame and fabric plane for permanent use.

### A Model Hydroaeroplane

The hydroaeroplane is in many ways far superior to the ordinary aeroplane, and it is very interesting to build a model of one of these machines and see it skitter along the water and rise from the surface.

Small model hydroaeroplanes are quite easy to make, and they have proved so successful and attractive that contests and meets for these tiny airboats are now a regular feature of model clubs.

Many of these models have flown 500 feet after running over the water but eight or ten feet.

The principal difference between a model hydroaeroplane and an ordinary model is in the "floats," or pontoons, as well as in slight details of proportion and dimensions to overcome the weight of the floats.

It must also be remembered that every part of a hydroaeroplane must be constructed of waterproof materials.

The fuselage consists of two pieces, each 32 inches long, of 1/4" x 5/16" spruce fastened together at the forward end in the usual way.

The cross-braces are heavier and stronger than in a strict flying-model, and are best set into I-beam section main beams, as described for the Japanese model.

The main plane is 17" x 3", built up of spruce ribs and bamboo edges, covered with bamboo-fiber paper on both sides, and coated with four coats of Ambroid varnish.

The floats are three in number, one in front and two in the rear, built of thin spruce and bamboo fiber.

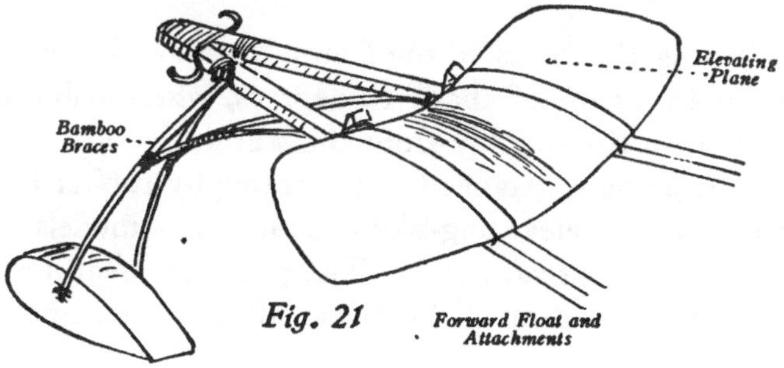


*Fig. 21*

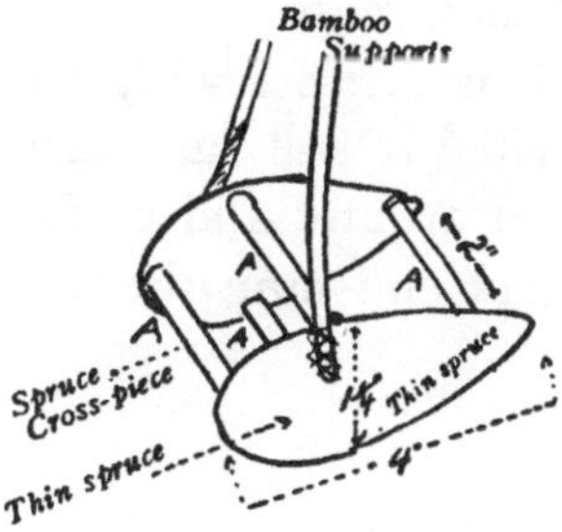


*Fig. 20*

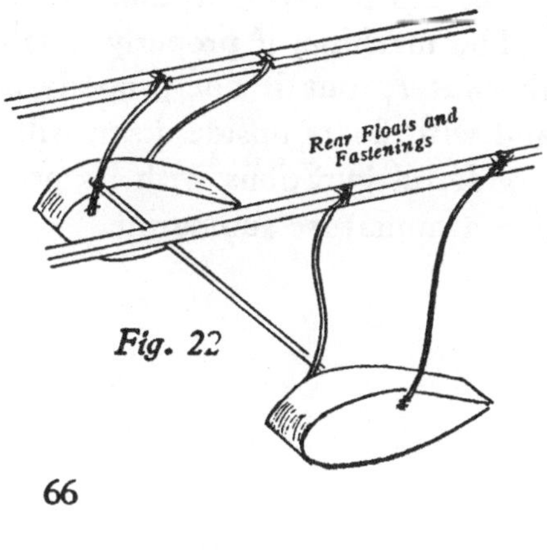


*Fig. 22*

The forward pontoon is 4″ long and 2″ wide and 1–1/4″ deep, constructed by cutting two pieces for the sides (Fig. 20), separated by four braces, as shown in Fig. 19, A, A, each 3/16″ square.

The floats are attached to the fuselage by sewing bamboo strips to the sides and lashing the ends to the main frame (Fig. 21). The rear floats are 2–1/2″ long, 1–1/8″ high, and 1–3/4″ wide, made in the same way as the front float, and with a cross-piece of bamboo connecting them, as shown in Fig. 22.

After sewing the floats to the bamboo supports they are covered with bamboo fiber and given four coats of Ambroid varnish.

In fastening the floats to the frame they should be set so as to have an incline of about 20 degrees, which will enable the machine to rise quickly from the water.

The front plane is attached to the frame by rubber bands, and rests upon an elevating-block, as in other models, with the hooks at the forward end. The propeller bearings and other details are exactly like those already described for racing-machines.

The propellers are 6–1/4″ in diameter, with a 10-inch pitch, and powerful strands are used as a motor.

The machine, if properly balanced, will rise readily from the water; but if not properly adjusted it will turn turtle, and with floats upside down will whirl around and perform wonderful gyrations, with its propellers churning the water like a miniature submarine.

# Chapter VI

## FLYING THE MODELS

ALL the racing and record-holding models, such as those already described, are more or less alike in form, construction, and details, and if you wish to be absolutely sure that your model will succeed and will fly it is wise to follow designs of this sort.

Many builders of real aeroplanes have had remarkable success in producing efficient machines by departing from established forms and building planes with oddly shaped wings, novel bodies, and other peculiar features.

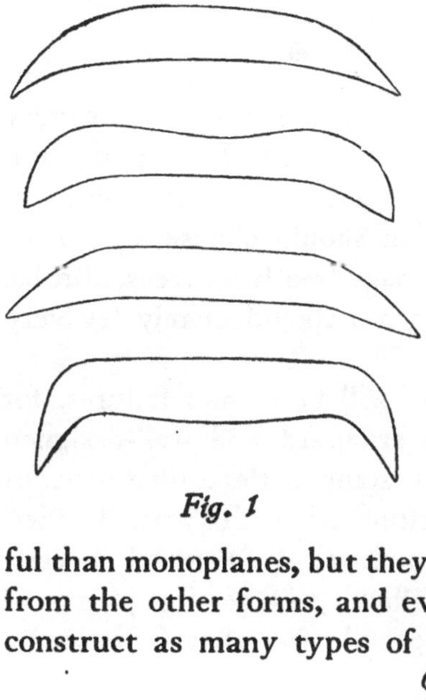

*Fig. 1*

In building or designing models the boy designer may also experiment and try wings and frames of various proportions and shapes.

Bird-like wings, such as those shown in Fig. 1, have proved most satisfactory on real aeroplanes, and for models they should also produce good results. Biplane models are as a rule less successful than monoplanes, but they serve as an interesting change from the other forms, and every boy builder should aim to construct as many types of models as possible.

By altering the shapes and sizes of planes and using one or two regular bodies a great deal may be learned in regard to the superior qualities of new planes, and when something is found that produces better results than others the machine may be laid aside as it is and considered a new design, while further experiments are carried on to improve upon it by the use of another model.

Sometimes a very slight alteration in the location or angles of planes, or in balance or proportions of the body, will produce unlooked-for and surprising results, and it is only by thus experimenting that new and faster models can be evolved.

It is in this experimental work that the greatest pleasure can be found, and once you learn the principles of model construction and learn to use tools on the minute parts required for models, you will find the work wonderfully fascinating.

### How to Fly a Model

After you have built your first model you will be very anxious to try it, and if you do not wish to be disappointed you should be careful in selecting the time and place in which to attempt a flight. You should choose a calm or nearly calm day and an open space free from trees, shrubs, buildings, or wires, and at first you should merely try very short flights.

Probably your first attempts will be dismal failures, for even if carefully made and a standard and well-designed model has been followed, yet some little adjustment or alteration will be necessary before the model actually flies.

Two people are usually required to fly a model successfully, but you will have no difficulty in getting plenty of your boy friends to help you; and after they have seen a few model flights they will all be anxious to build models of their own, and very soon you will be able to form a club and hold regular contests.

Various clubs are already formed in most of the larger cities, but it is lots of fun to have your own little club; and after you have built a number of models and have learned to fly them well you may be able to arrange competitions with other clubs just as you would form a baseball nine and, after practising, challenge other nines to games.

Model aeroplanes are now so firmly established, and have become so common, that regular rules and regulations for contests and meets have been formulated and are accepted by nearly all clubs, just as baseball or football games are subject to recognized rules which all players must abide by.

Of course a great many contests and meets are unofficial, or are held just for fun or for practice; but oftentimes wonderfully long or fast flights are made at these meets, and, unfortunately, such flights do not go on record as official.

The principal rules for model-aeroplane contests, as compiled by Mr. Edward Durant, are as follows:

1. A contest to be official must have at least five contestants.
2. Each contestant must abide by the rules of the contest and the decision of the judges.
3. Each contestant must register his name, age, and address before the event.
4. Each contestant must enter and fly models made by himself only.
5. Trials to start from a given point indicated by the starter of the trials, and distance is to be measured in a straight line from the starting-point to where the model first touches the ground, regardless of the curves or circles it may have made.
6. Each contestant must have his models marked with his name and number of his models (1, 2, 3, etc.), and each model will be entitled to three official trials. Contestant has the privilege of changing the planes and propellers as he may see fit, but only three frames can be used in any contest. If in the opinion of the board of judges there are too many entries to give each one nine flights in

the length of time fixed, the judges have the power to change that part of rule No. 6 to the following:

"Six flights will be allowed to each contestant, which can all be made with one model or divided up; due notice must be given to each contestant of the change."

7. No trial is considered as official unless the model flies over 100 feet from the starting-point. (The qualifying distance can be changed by agreement between the club and the starter provided the entrants are notified.)
8. Should the rubber become detached from the model, or the propeller drop off during the trial, the trial is counted as official, providing the model has covered the qualifying distance.
9. Contests should cover a period of three hours, unless otherwise agreed.
10. The officials should be: a starter, measurer, judge, and scorer; also three or four guards to keep starting-point and course clear. The first three officials shall, as a board of judges, decide all questions and disputes.
11. A space 25 feet square (with stakes and ropes) should be measured off for officials and contestants, together with one assistant for each contestant. All others must be kept out by the guards and a space kept clear (at least 25 feet) in front of the starting-point, so a contestant will not be impeded in making his trial.
12. Each official should wear a badge, ribbon, or arm-band designating his office, and must be upheld in his duties.

### HANDICAPS

13. At the discretion of the club there may be imposed a handicap for club events as follows:

A contestant in order to win must exceed his last record with which he won a prize.

### COMBINATION AND DURATION EVENTS

14. First, second, and third records to count. Lowest number of points to win. For example:

A—may have 1st in distance and 2d in duration=3 total points
B—may have 3d in distance and 1st in duration=4 total points
C—may have 2d in distance and 3d in duration=5 total points

Thus A wins.

### RISING FROM THE GROUND

15. Models to be set on the ground and allowed to start off without any push from the contestant. Models must rise from the ground to clear a rope or stick laid on the ground, and no flight is to be con-

sidered as official unless it does so. Contestant may start at any length back from the mark he chooses, but the distance is to be measured only from the rope or stick.

OPEN EVENTS

16. These events are open to all, and no handicaps are imposed on club members or others.

## RULES FOR INTER-CLUB MODEL AEROPLANE TOURNAMENTS

*(For a Club Trophy)*

Compiled by Mr. Edward Durant

The tournament to consist of five events as follows:

DURATION: Models launched from hand
DISTANCE: Models launched from hand
DURATION: Models launched from ground
DISTANCE: Models launched from ground
DURATION: Models launched from water

Arrange dates about two weeks apart, to allow time for special model to be made for the next event.

In case of inclement weather, the event to take place the week following, so that the regular schedule for bi-weekly events can be carried out.

Each competing club must be represented by a team of three contestants, and one non-competitor, who will act as judge in conjunction with the judges from the other clubs, and a manager selected by the judges who will issue calls for meetings.

There shall be a meeting of the judges of the competing clubs at some designated headquarters, at which time the dates and general details shall be arranged, and between events there should be a meeting called, for general discussion regarding the last previous event, receive protests and suggestions, and to announce officially the result of same.

The manager shall have control of the various events, assisted by the judges, and they shall decide all disputes that may arise, and act as scorers and timers as well.

Each flyer will be allowed but one model and shall be entitled to three official flights, but he shall be permitted to make any repairs or replace any broken parts. Each event shall close when all the contestants have made three official flights, or when three hours' time has elapsed.

To determine results of events, the best flight of each member of the team shall be added together, and the total divided by three. The team having the largest result shall be declared winner of that event, and shall be credited with 20 points. (So in case one club wins all five events it will have 100 points.)

The other teams getting points in proportion as their average is to the winner, and, should some member or members not make an official flight, the record must be figured in the result as O, for instance:

| | | |
|---|---|---|
| A flies | 1,500 | feet, his greatest flight |
| B flies | 0 | No official flight |
| C flies | 600 | feet, his greatest flight |
| | 2,100 | feet |
| 1/3 = | 700 | feet, the average of the team |

At the close of the tournament a meeting shall be called by the manager to determine the net result by examining the official score for each event, and figuring out the proportion due to each club.

In determining what will qualify as an official flight, judgment must be used by the manager and board of judges, not to set the qualifying marks higher than the capabilities of the average flier of the clubs that will go into the contest; for example:

There has recently been held (May and June, 1913) a five-event inter-club tournament in which the qualifying points were as follows:

| | | |
|---|---|---|
| DURATION: | Launching from the hand, | 60 seconds |
| DISTANCE: | Launching from the hand, | 1,200 feet |
| DURATION: | Launching from the ground, | 30 seconds |
| DISTANCE: | Launching from the ground, | 600 feet |
| DURATION: | Launching from the water, | No limit |

These high marks were set because the contestants were in the world's championship class, comprising such fliers as Selley and Herzog, of New York M. A. C.; Cavanaugh, of Long Island M. A. C.; Myers, of Summit, New Jersey, M. A. C.; and the Bamberger brothers, of Bay Ridge M. A. C.; and the results justified the high standard, as the following three new world's records were made:

| | | |
|---|---|---|
| W. Bamberger | Making duration from ground . | 81 seconds |
| L. Bamberger | Making distance from ground . | 1,542 feet |
| Geo. H. Cavanaugh | Making duration from water . | 60 seconds |

But for the gusty winds which prevailed on the occasion of the duration from hand and distance from hand, new world's records would probably have been made for these events.

While the list of events here used covers all styles of flying, still the number of them can be cut down or modified to suit the qualification of the clubs which may compete.

These are special rules for inter-club series of events. The general rules for model aeroplane contests will prevail where they do not conflict with the above special rules.

### Measuring Flights

Formerly all flights were measured by a tape or similar method, but with the increase in model-aeroplane flying better appliances were produced, and nowadays all flights are measured by automatic instruments that are simply wheeled over the ground.

Some of these measuring-devices are very accurate and are also very costly; but any boy with a little ingenuity can make a measurer that will serve every purpose and will prove very accurate—far more accurate, in fact, than any tape or rule. The following device was originated by Mr. Edward Durant, and will prove very accurate and useful.

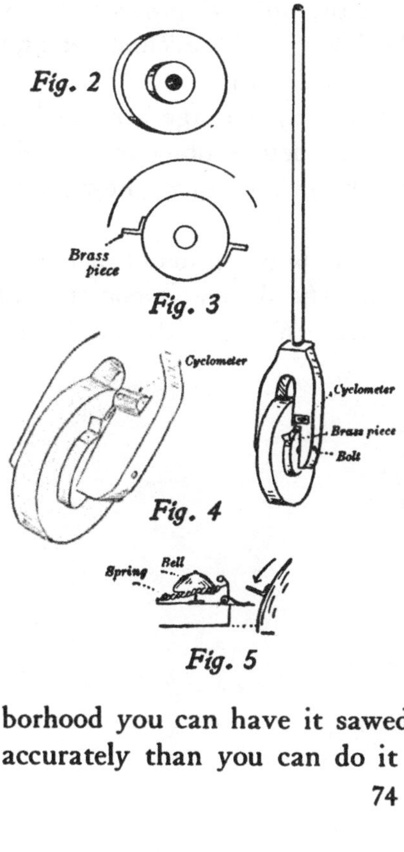

*Fig. 2*
*Fig. 3*
*Fig. 4*
*Fig. 5*

### How to Make a Measuring-Device

First secure a wheel exactly two feet in circumference. An old velocipede, bicycle, or baby-carriage wheel will answer; or make a wooden wheel by nailing boards together with their grain running in opposite directions, and saw it out round and true. If there is a sawmill or good wood-working shop in your neighborhood you can have it sawed on a band-saw much more accurately than you can do it yourself.

If you use such a wooden wheel you should cover its edge with brass or iron as a tire to protect it.

To one side of this wheel fasten a small wheel three inches in diameter (Fig. 2), and on the edge of this small wheel fasten two pieces of brass or iron exactly opposite each other (Fig. 3).

Saw out a piece of plank or board in the shape shown in Fig. 4, and fasten the wheel in the slot by a bolt for an axle, as shown in the cut, and attach a handle at the upper end.

A cyclometer that registers feet, or that plainly registers the number of revolutions, should be fastened near the upper end of the frame, so that the projecting bits of brass on the small wheel strike the lever of the cyclometer and thus register the distance the wheel turns.

Instead of the cyclometer you may arrange a small bell with a striker, and as you push the device along you merely have to count the strokes of the bell to know how many feet you have traveled.

Both of these arrangements are shown in detail in Fig. 5, and you can easily see just how the device is constructed and how simple it is to measure distance with this home-made apparatus.

# Part III

# GLIDERS, OR NON-PROPELLED AEROPLANES

## Chapter VII

### TYPES OF GLIDERS

WHEN you have learned something about the principle of aeroplanes and have built model machines you will no doubt long to try real flying.

Real man-carrying aeroplanes are hard to build, and are expensive, while the motors required to drive them are far beyond the reach of most boys.

There is, however, a way to enjoy all the pleasurable sensations of actual flight in a simple and inexpensive way, and that is by the use of a glider.

A glider is a form of aeroplane which has no propelling power or motor, but depends upon a wind or a jump from a height to carry it forward in flight.

Gliders are really very important things, for the various experiments that led to the invention and construction of the first real aeroplanes were made by the use of gliders, and had it not been for these safe and simple things probably aeroplanes would never have been invented.

In a way gliders are merely aeroplanes without a motor; and, as no engine or steering-apparatus is required and no propellers are used, the glider may be constructed far lighter and more easily than a real aeroplane.

There are a great many forms of gliders, for, like the aeroplane-builders, each designer of a glider tried his best to invent something better and safer than others; and, while many proved failures, a great many others have proved so very successful that they are recognized as true types.

Some of the best known are the Montgomery glider, the Chanute glider, the Wright glider, the Witteman glider, and many others.

As gliders are the ancestors, so to speak, of aeroplanes, they are very similar in form to the more highly perfected machines; and we will find monoplane gliders, biplane gliders, and triplane gliders, as well as a few multiplane gliders, which have a great many small planes instead of the one, two, or three large planes of the other forms (Fig. 1).

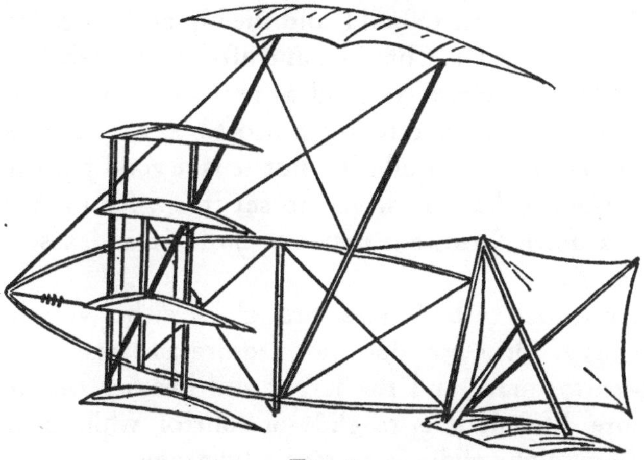

*Fig. 1*

Aeroplanes are very dangerous machines in the hands of inexperienced people, and even in the hands of expert airmen accidents often happen and many aviators are killed.

Gliders, on the other hand, are very safe, for unless you glide from a great height or in a gale of wind there is little danger of the slightest harm coming to you even if you fall or upset, and about the worst that can happen is to break your machine or get a few bruises or torn clothes.

In a way gliders are even more like kites than aeroplanes, for they rise and sail against the wind; and, while at times they glide for quite a long distance and actually rise quite a good deal higher than the spot from which they started, yet as a rule the flight of a glider is comparatively short, and is a long, descending slide through the air.

It requires but little skill to guide a glider, for if properly built and balanced they keep right side up and sail steadily with little effort on the part of the passenger.

Of all types of gliders the biplane is perhaps the safest as well as the easiest to build, and the biplane glider invented and perfected by Mr. Chanute, and afterward slightly altered to the Wright style, is as good as any to commence on.

The cost should not be over $10 to $15, and no special tools are required to build it; but it is a good plan to have a large shed or barn in which to set it up, for even if built in knock-down form it is easier to build it and assemble it ready to fly.

It is not wise to build your first glider either very large or very heavy, for a boy does not require as large a glider as a full-grown man; and the larger and heavier the machine the more difficult it is to glide or control, while in making a landing a light glider is a great advantage.

The dimensions should be closely followed, however, for a slight variation may mean total failure, and if you wish to experiment with alterations and innovations it is wise to wait until you have built at least one successful machine of known strength and ability to fly.

After you have done this you can build another with any changes you may think will improve it, or you may try your hand at monoplane or triplane affairs.

Before commencing to build a glider, or even to get the plans or materials ready, you should study the principles and operation of these machines, for you will succeed far better if you know beforehand just what purpose each part is for, and how it affects the whole.

### Principles of Gliding

Just as the aeroplane flies by being driven against the air by its motor and propeller, so the glider flies by being forced into the air by the weight of the operator's body having a tendency to fall.

You cannot run with your glider and lift and fly; but if a glider is towed rapidly behind a horse or an automobile it will rise and fly through the air as long as the forward motion continues.

This is not, however, true gliding, and in order to glide in the proper manner you must jump into the air while running downhill; and, although the same result may be accomplished by jumping off a building, the latter method is excessively dangerous, and should never be attempted.

You should always run toward the wind when trying to glide, but you should *not* try to use a glider in a strong, gusty, or uncertain wind.

Never try to glide over broken, uneven, rough, or rocky ground, but select a smooth, soft slope free from stumps, sticks, brush, or other objects that may injure you or the glider.

Even if these things are not in your path, they may influence the air currents sufficiently to make flight uncertain or difficult.

Few people realize how much a little thing will influence the air, especially on a windy day.

Photo by Frank Wiehn

A GLIDER IN FLIGHT, OAKWOOD HEIGHTS, STATEN ISLAND, N. Y.

Close to the earth the atmosphere is composed of numerous swirling and confused currents, which constantly rise or fall, and by watching particles of dust or dead leaves on a fairly windy day you can see this very easily.

On almost calm days these currents still occur, but are far less pronounced, and even in perfectly calm weather rising and falling currents are caused by various temperature changes, such as the sun passing behind a cloud, or by other causes which we cannot see or feel.

As cold air is heavier than warm air, the latter is constantly rising from the earth while the sun shines and warms it, while the cold air is descending and filling the space left by the warm air.

The more unequal the temperature the more these currents rise and fall, and even at great heights these varying, unseen air streams occur and prove the most dangerous items in flying.

Even when the wind is very steady a small object, such as a tree or building, will cause the air currents to divide, and the least disturbance thus caused will affect the other air currents for a long distance in every direction.

As a rule, however, the cross-currents and miniature whirlwinds caused by variations in temperature or by objects in the path of the wind, are not great enough to disturb a glider unless the wind is blowing more than fifteen miles an hour; but it is far wiser to try your first few glides in a very light wind.

Begin by running against the wind on level ground and making short jumps. As soon as you have learned the sensation of actually traveling a short distance in the air, and have discovered that you can land on your feet and can prevent the machine from capsizing, you may attempt a real glide from a low hill.

### Making a Glide

Carry your glider to the top of the hill, and if possible have two of your friends hold the ends of the lower plane.

Get beneath the machine and grasp the front main beam and lift the glider until the arm-sticks are tight up beneath your armpits (Fig. 2).

Now run quickly down the slope, your friends letting go of the ends as soon as you are in motion, and you will at

*Fig. 2*

once find that the machine has apparently lost all weight and is actually trying to lift you from your feet.

Now elevate the front edge of the glider slightly and leap into the air. If your glider is properly made and your position is right you will sail smoothly and evenly off into the air and gradually sail down to the foot of the hill. As you approach the earth push yourself toward the rear of the glider, so that the machine tilts up. This will cause it to rise a little; but almost instantly it will lose its forward motion, and you will slowly and gently drop to the ground on your feet.

If in your glide you find the machine has a tendency to droop toward one side or the other, you merely have to swing your legs and body toward the higher side, and the machine will come back to a level keel.

Only a very slight motion is required to accomplish this lateral balance, and at first you will probably err by swinging too far or too quickly.

If you swing the body forward the center of gravity will be brought toward the front, and the machine will descend, while by swinging the weight farther back the machine will tilt up in front and will lose its speed and drop you to earth.

As a rule, the tendency of beginners is to place their weight too far back; but after a little practice the operator will learn how to handle his weight so as to increase the velocity or decrease it at will.

The diagram (Fig. 3) shows how the weight of the operator may be shifted to bring about any desired line of flight.

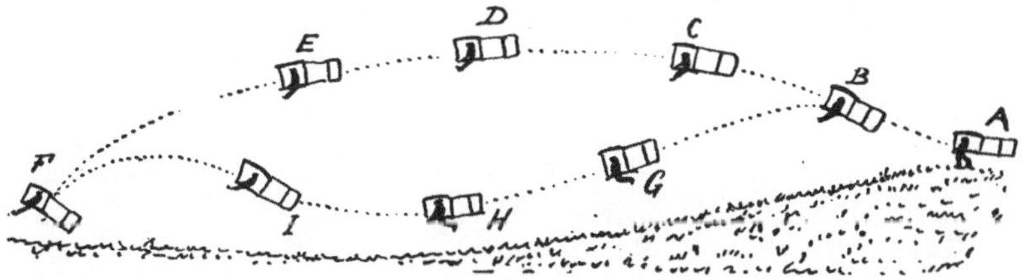


*Fig. 3*

At A the flier is just starting to run forward on the hilltop, and the line from A to B shows how his flight or glide starts at first.

If the weight is too far back the glider will continue as in the line to F through C, D, E. If at the position B the operator moves his legs back, as shown, the glider will travel more sharply downward, as shown by the line from

B to G, H. If at the point H, where the machine is traveling the fastest, the legs are swung forward again, as at I, the machine will rise slightly and will finally settle to the earth at F.

This sort of a glide is the most enjoyable as well as the safest, for a longer flight is made at a lower elevation and with less likelihood of a bad fall.

Always remember that a gust of wind from the *front* will cause the glider to *rise*, and a gust from the *rear* will cause it to *drop*. With these few hints in regard to gliding, I will tell you how to build the glider, and if you follow the directions carefully you will have no trouble in making a flight the first time you try.

## Chapter VIII

### HOW TO BUILD A GLIDER

BEFORE commencing to plan or build any form of glider it is best to be thoroughly familiar with the names of the various parts, in order more clearly to understand the directions.

A glider is composed of comparatively few parts, and these are clearly shown in the diagram (Fig. 1).

This illustration shows the glider frame complete; but the ribs on the lower plane and the fabric covering of the two

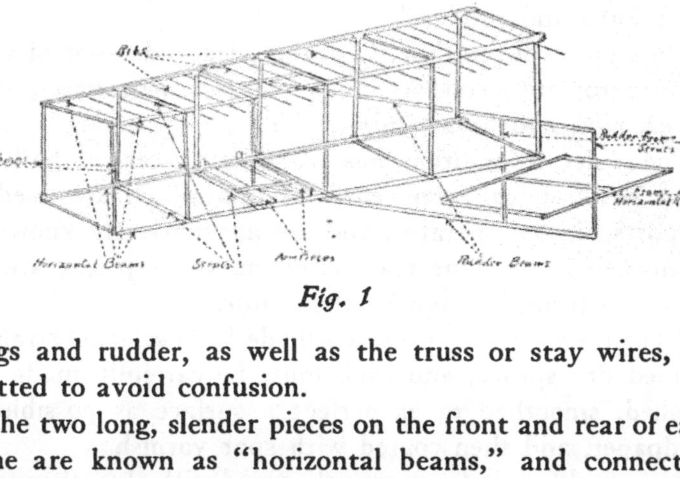


*Fig. 1*

wings and rudder, as well as the truss or stay wires, are omitted to avoid confusion.

The two long, slender pieces on the front and rear of each plane are known as "horizontal beams," and connecting the upper and lower horizontal beams are a number of upright pieces known as "stanchions." Connecting the horizontal beams of each plane from front to back are six pieces in each plane, and these are called "struts."

Crossing the planes from front to rear between the struts are numerous smaller, curved pieces known as "ribs." Between the two central struts of each plane, and placed parallel with the horizontal beams, are short, stout pieces to which the long "rudder-beams" are attached, and the latter are connected at their extreme ends and at another spot between their ends and the planes by braces, or struts.

Between these two "rudder-frame struts" a rectangular frame is placed, which when covered with fabric will form the "horizontal rudder."

Thus you will see that the frame of a glider—or of a biplane aeroplane—consists of *horizontal beams*, *struts*, *stanchions*, *ribs*, and *rudder-frames*.

The glider, in addition, has two stout pieces on the lower plane between the two central struts which are used as supports for the operator, and are appropriately known as "arm-pieces." In the real aeroplane these pieces are replaced by beams to support the motor.

All of these various pieces are made from selected straight-grained dry spruce, and each must be carefully made and finished, smoothed to as perfect a surface as possible by sandpaper, and then coated with spar varnish.

Some builders prefer to coat the wood with glue, then sandpaper it smooth, and finally finish with shellac, but the varnish is easier to use and answers every purpose.

It is usually a difficult matter to procure straight-grained spruce of sufficient lengths in an ordinary lumber-yard; but if you live near any city where aeroplanes are built or are used you can secure the material from the builders of the aeroplanes, or from firms making a specialty of this kind of lumber.

If no long pieces can be procured for the main beams, two shorter pieces can be spliced together.

## Quantity and Sizes of Material

*For the horizontal beams you will require:*
Four pieces of spruce each 20 feet long, 1–1/2" wide, 3/4" thick.
*For the struts:*
Twelve pieces, each 3 feet long, 1–1/4" wide, 1/2" thick.
*For the stanchions:*
Twelve pieces, each 4 feet long and 7/8" square.
*For the ribs:*
Forty-one pieces, each 4 feet long, 1/2" square.
*For the arm-pieces:*
Two pieces, each 3 feet long, 1" wide, 1–3/4" thick.
*For the rudder-frame:*
Two pieces 8 feet 11 inches long, two pieces 3 feet 10 inches long, four pieces 2 feet long, and two pieces 6 feet long. All of these 1" square.
*For the rudder-beam supports:*
Two pieces 2 feet 11–1/4" wide and 3/4" thick.
*For covering the planes:*
About 20 yards of fabric, either silk or cotton, 1 yard wide. Stout silk or unbleached cotton will answer, but it is better to buy the aeroplane fabric, for sale by all dealers in aeroplane supplies, as this is very tough and light, and is already treated to render it air and water tight.
*For stays you will need:*
A roll of No. 12 piano-wire.
*You will also require:*
Twenty-four strut or stanchion sockets. These may be purchased ready-made, and, as they should be strong and reliable, the ready-made sockets are far better than anything you can make yourself or have made to order. For the rest, the only things needed are copper or galvanized-iron carpet-tacks, some stove-bolts, shellac, varnish, glue, and simple carpenter's tools.

## Preparing the Materials

Having cut out the various pieces of wood and smoothed them down to the dimensions given, you should proceed to round off all the corners, and on the stanchions work them into an elliptical or stream-line shape (Fig. 2).

The pieces for the ribs should then be steamed and bent on a form made of wood, and curved so as to bring the central part of each rib about two inches higher than the ends (Fig. 3).

By making a form of planks and nailing thin strips across (Fig. 4) a number of ribs may be readily bent at one time, thus insuring a uniform curve and making the work much easier.

The curve required is so slight that little trouble will be encountered in bending them, and they may easily be held

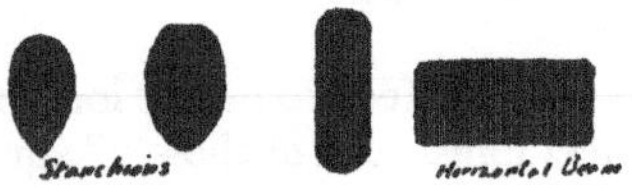

*Fig. 2*

*Fig. 3*

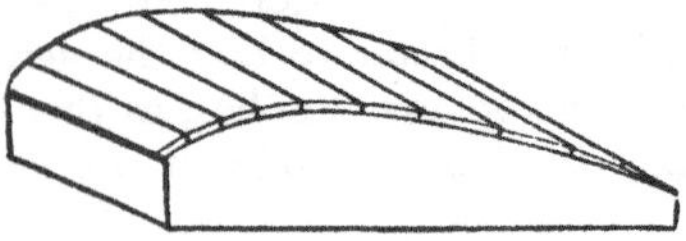

*Fig. 4*

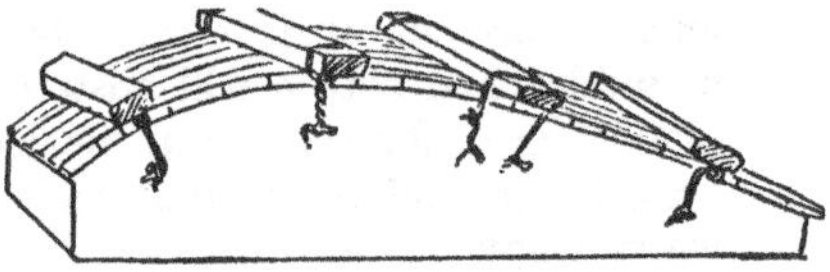

*Fig. 5*

in position while drying by cleats placed across them and held down to the sides of the form by a hook or by twisted wire, as shown in Fig. 5.

Although steaming is the best way to prepare the ribs for bending, yet soaking in boiling water will answer just as well, and for this purpose an old wash-boiler will serve very well indeed.

After the ribs are steamed and bent, and all the pieces are rounded off on the corners, and the stanchions are worked down to the proper shape, they should all be sandpapered just as smooth as possible and then varnished. When the first coat of varnish is thoroughly dry it should be smoothed off with fine sandpaper and a second coat applied. When this second coat is dry and hard you can proceed to put the glider together.

### Assembling the Glider

Select two of the long horizontal beams, and place them on a smooth, level floor so they are three feet apart and parallel.

Between these place six of the strut pieces, one at each end and the others spaced 4–1/2 feet distant.

This will give you a rectangle, formed by the two horizontal beams and the two end struts, and divided up into five sections or cells with the two central struts two feet apart and the others 4–1/2 feet apart (Fig. 6).

Be very careful in placing the struts and beams to see that the ends of the struts are snugly fitted against the beams and that each point of contact is an exact right angle.

Test the corners repeatedly with a steel square, for if the plane skews or twists you cannot possibly make a good flight with the machine.

Having made sure that the struts all fit the beams perfectly, you should then mark the position of each, and carefully cut away one edge of each end of each strut for a distance of 1–1/2" from the end and 3/4" deep (Fig. 7).

When the notches are cut square and even place the struts on the beams and fasten them in position by small wire-nails and glue.

As soon as the glue has set the struts should be fastened to the beams by means of "clamps," which are merely pieces of sheet brass 1/16" thick, 3–7/8" long, and 1" wide, with the ends rounded and a 1/4" hole bored in each end. You can readily cut out these clamps yourself; but, as they are

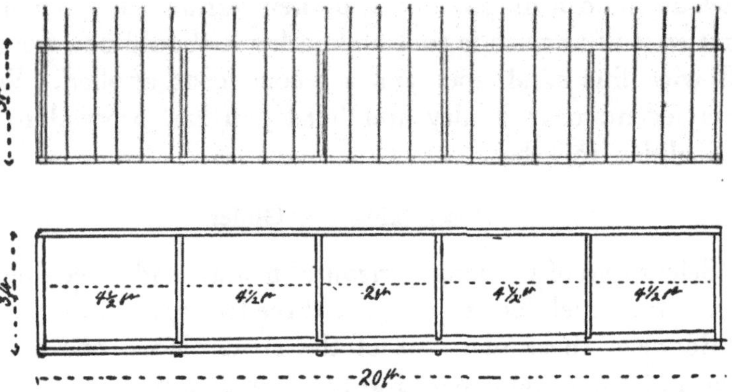


*Fig. 6*

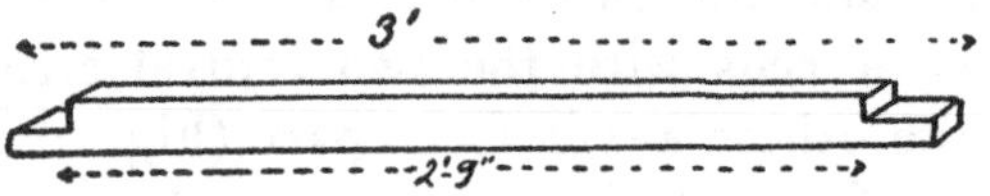


*Fig. 7*

*Fig. 8*

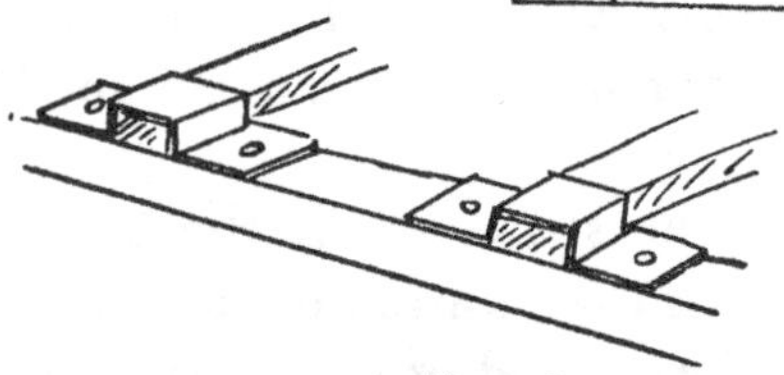

*Fig. 9*

to be held in position by the stanchion-socket eyebolts, you need not be in a hurry about them, but can make them up while waiting for the glue to dry, or at odd times.

As soon as you have completed the beams and struts for one plane you may finish the other in the same way.

Next place one frame with the *struts on top*, and the other with the *struts beneath*, and on these place the ribs.

The ribs should be placed with the curved side up (Fig. 8);

and on the upper plane—the one with the struts on top—there should be twenty-one ribs placed one foot apart; and on the lower plane—the one with the struts underneath—twenty ribs are placed, leaving a space in the middle two feet wide for the operator's body.

When the ribs are placed in position and perfectly square with the beams, and with one end flush with the latter, and the other end projecting one foot beyond the beam (Fig. 9), they may be fastened in position.

First tack them in place with fine wire-nails, and then place over each a strip of sheet copper or brass 2-1/4″ long and 5/8″ wide with the ends rounded, and fasten the ends to the beam on each side of the rib with No. 5 round-headed brass wood-screws, about 1/2″ long.

In using these screws be sure and start a hole with a fine awl, as otherwise you will be very liable to crack or split the wood, and if this occurs the piece should be discarded and a new piece used.

### Connecting the Planes

The two frames of the planes, or wings, having now been completed, and the ribs clamped in position, you can proceed to fasten them together by the stanchions.

It is in this work that you will require the twenty-four metal sockets of aluminum or brass.

These sockets are of the form illustrated in Fig. 10, and may be purchased for a few cents each from any dealer in aeronautical supplies.

The sockets are made to receive either oval, stream-line, or round stanchions, and are of various sizes, to suit machines of different kinds.

The dimensions of those for the glider should be about as follows:

Base, 3-1/4" long, 1-1/4" wide, 1/4" thick.

Internal diameter of cup, 7/8"; outside diameter, 1-1/4".

Height from base to top of cup, one inch.

Holes in base, 1/4" diameter, separated 1-7/8" apart.

Two smaller holes 3/16" in diameter should be bored in the base outside of the larger holes.

Having the sockets ready, place six of them on the front beam of the plane, beginning them at the end of the beam and placing one socket exactly over the end of each strut on the opposite side of the beam, as shown in Fig. 11.

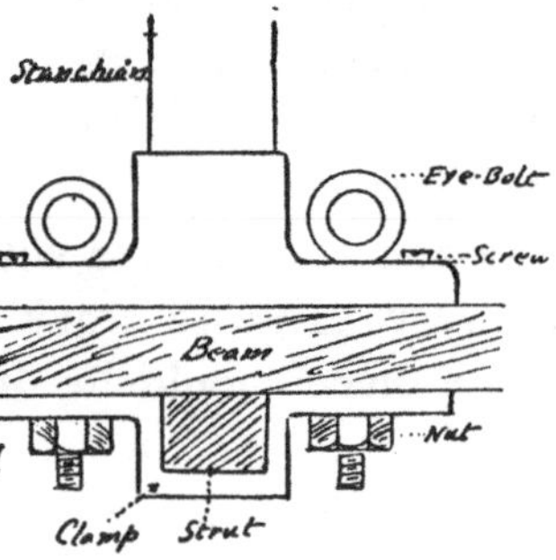


*Fig. 11*

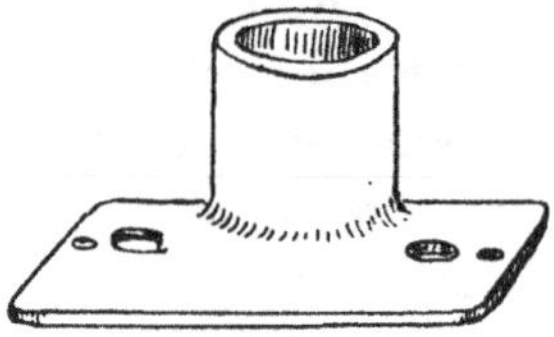

*Fig. 10*

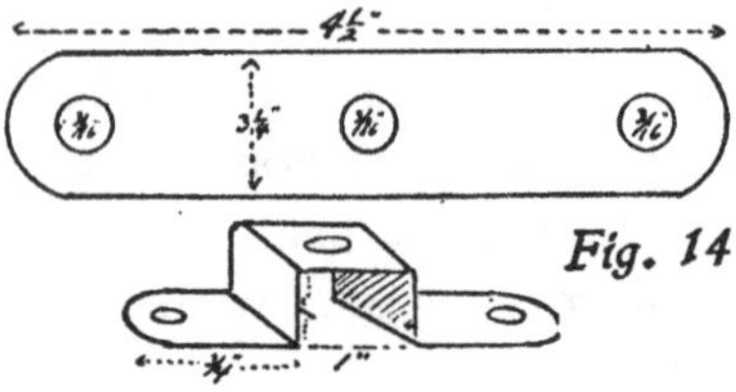

*Fig. 14*

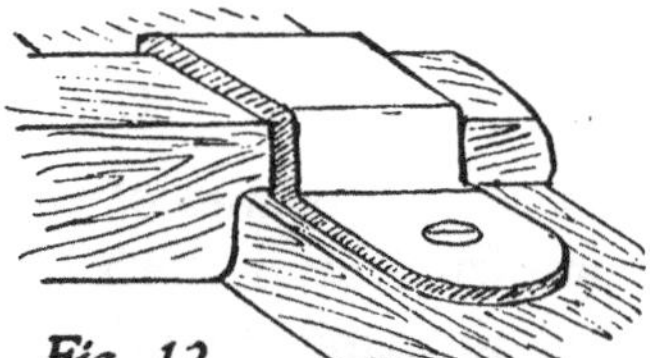

*Fig. 12*

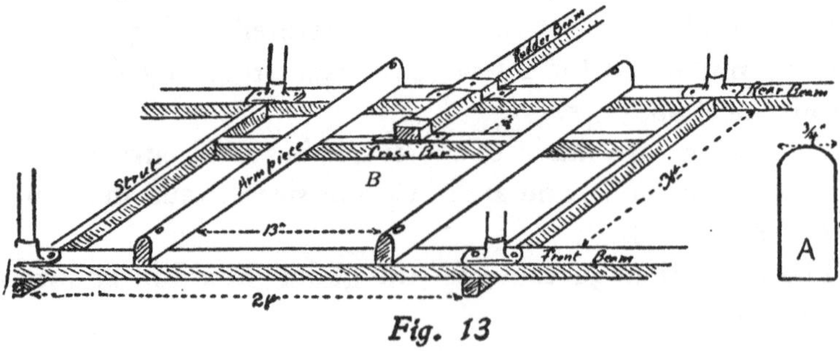


*Fig. 13*

Screw each socket in place with small wood-screws with round heads through the smaller holes in the base.

Next place the brass clamps you have made for the struts over each strut, and pass the eyebolt of the socket down through the base of the latter and through the 1/4″ holes in the clamp beneath, and screw on the nut so as to firmly clamp the socket, beam, strut, and clamp together (Fig. 12).

Repeat the placing of another set of six sockets on the rear beam of the lower-plane frame, and when all these are in position treat the upper frame in the same way, being sure that the struts of the upper frame are *above* the beam, while those of the lower plane are *below*.

The next step is to set up stanchions in the lower-plane sockets, and in doing this you should use a great deal of care to keep the ends snug in the cups.

If they are too tight to slip in they should be rubbed down slightly with fine sandpaper, but under no consideration should you whittle or plane them down or try to drive them into place.

Having all the stanchions set up in the sockets of the lower plane, you must now lift the upper-plane frame and set it on the stanchions.

In doing this you will probably need two friends to help you, for the frame is clumsy and ungainly even though light in weight, and it is a difficult matter to handle it alone and place the sockets over the upper ends of the stanchions.

If two of your friends will lift the frame and support it in position, it is very easy to fit each stanchion into its proper socket in turn and get them all evenly and tightly in place.

The next step is to place the arm-pieces in position, which is very easily accomplished by boring 3/16″ holes through the arm-piece and the lower-frame beams, and bolting the pieces in place 6–1/2″ on each side of the exact center of the beams, so that a space 13 inches wide is left between the arm-pieces (Fig. 13).

The arm-pieces should be rounded smoothly on top, so as to be flat on the bottom and half round above, as shown in the section A, Fig. 13. Eight inches from the rear lower beam a cross-piece 2′ 11–1/4″ long and 1–1/4″ x 3/4″ should be fastened between the two central struts, and another placed in exactly the same position on the upper frame (Fig. 13, B).

These are fastened in position by glue and fine wire-nails, and are further secured by small angle-irons of brass, aluminum, or steel screwed onto the struts and cross-pieces by round-headed screws.

These angle-irons may be made from sheet brass or aluminum, or may be purchased ready-made.

These cross-pieces are designed to support the rudder-frame, and you should now construct the latter and some sockets for fastening it in place.

The rudder-frame sockets are very simple affairs of sheet brass 1/16″ thick and 4–1/2″ long. Two of the sockets are 3/4″ wide and two 1–1/4″ wide, and are bent as shown in Fig. 14. The ends of each are rounded, and are drilled with 3/16″ holes at the ends, as shown in the cut, while the smaller ones also have a hole through the center, as illustrated.

One of the larger sockets is fastened to the main rear beams, and the smaller ones to the cross-pieces by bolts as shown in Fig. 13, and the utmost care must be taken to place these sockets absolutely in line and exactly in the middle of the planes.

The frames should now be quite stiff and strong, but they must be still further strengthened by trussing and staying with wire, in order to support the weight of your body and bear the strain and force of the air against their surfaces.

## Chapter IX

### COMPLETING THE GLIDER

TO properly truss the frames the framework should be supported on benches or horses, so as to be perfectly level and true without any strain coming on any one part.

The roll of No. 12 piano-wire should then be placed close at hand with pliers, some short pieces of copper or brass tube 1/8″ inside diameter, and about 3/8″ long, and soldering-tools.

There are two methods of trussing, either one of which is excellent for a glider; but the first, while much the simpler, is also somewhat weaker.

For aeroplane use the simple method is seldom used, but for glider construction it serves very well; and, as it answers every purpose and is much easier to describe and understand, it is the only method you need consider at present.

In many gliders the builders depend upon pulling taut the wires by hand, but this is a poor and difficult method, and it is far better to use small turnbuckles.

Turnbuckles may be made from old bicycle spokes; but these are apt to be weak and unreliable, and as first-class and standard turnbuckles may be bought of dealers for less than fifteen cents each, it is not worth while to trust your life and the glider to home-made makeshifts.

To truss a section of the frame slip a piece of the copper tube over the end of the wire, pass the wire through an eyebolt in a socket, run the end back through the copper tube, bend the end of the wire back over the tube, and shove the tube down close to the eyebolt, as shown in Fig. 1. Some builders trust to this fastening entirely, but I advise all builders to solder the wire and tube firmly together to insure greater strength and safety.

*Fig. 1*

The soldering can, however, wait until all wires are in position, and in this way time will be saved and a better job accomplished.

Having fastened one end of a wire to an eyebolt, place a turnbuckle on the opposite diagonal corner and loosen it

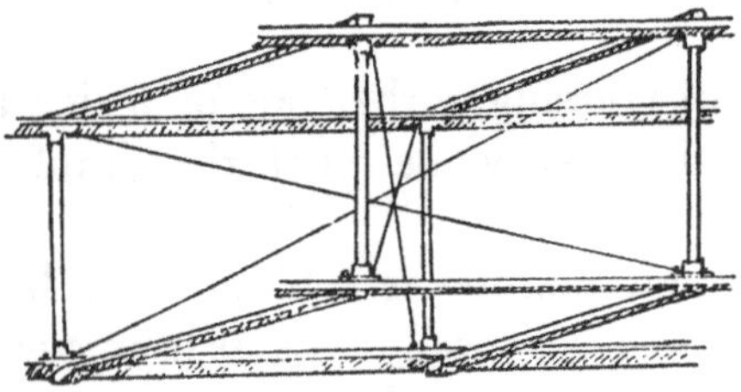

*Fig. 2*

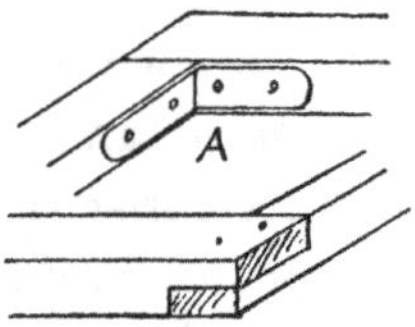


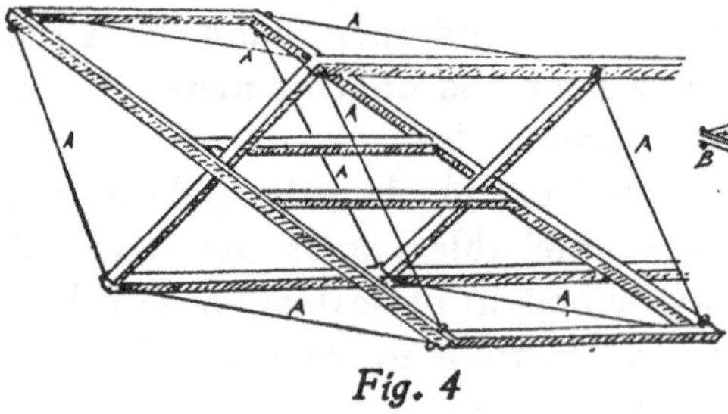


*Fig. 4*

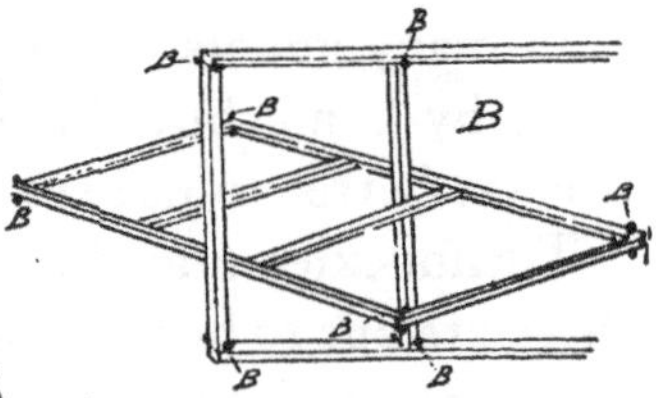


*Fig. 3*

out nearly its full extent. Stretch the wire down to the eye in the turnbuckle and fasten the end through it exactly as you fastened the other end to the eyebolt, and pull it taut as possible with your hands before fastening.

Repeat the operation of placing the wire and turnbuckles on the other corners of the section, and the wires will then appear as in Fig. 2.

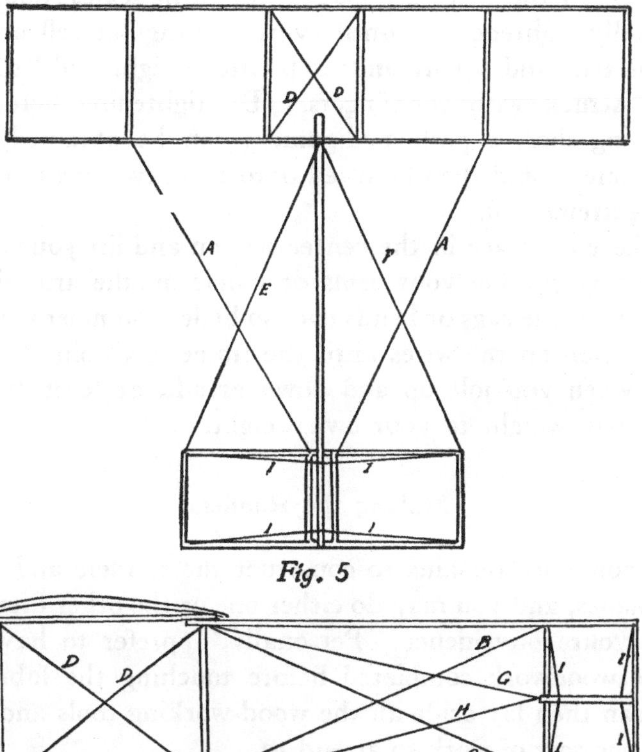


*Fig. 5*

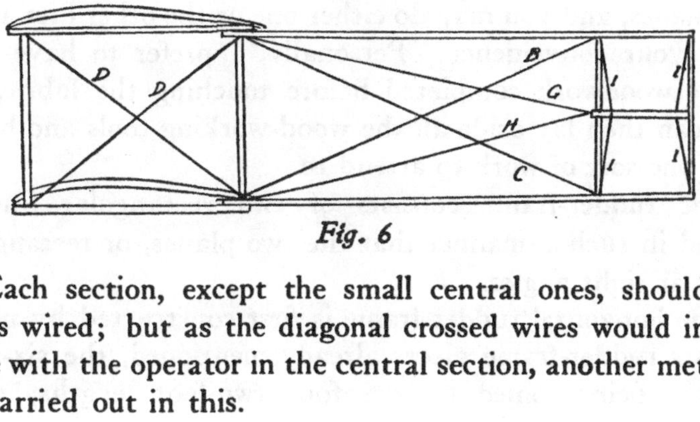


*Fig. 6*

Each section, except the small central ones, should be thus wired; but as the diagonal crossed wires would interfere with the operator in the central section, another method is carried out in this.

Here a wire is run diagonally from the junction of each center strut with the upper beams to the opposite corner (Fig. 6, D), and other wires are run diagonally across from the corners of the rear stanchions with the rear beams, as shown in Fig. 5, D.

When all the wires are placed, the turnbuckles should be gradually tightened up until every rectangular cell or section is true and square and each wire is tight and "sings" when struck with the fingers. By tightening here and loosening there a perfect alignment can be obtained, and the frame should then be lifted onto two saw-horses, one at each extreme end.

Take your place in the center section and lift your entire weight by placing your arms or hands on the arm-pieces. If the machine sags or bends ever so little, you must increase the tension on the wires until the frame is absolutely rigid even when you jolt up and down or add at least twenty pounds in weight to your own weight.

### Making the Rudders

It now only remains to construct the rudders and cover the planes, and you may do either one or the other first as it suits your convenience. Personally, I prefer to have the entire woodwork completed before touching the fabric, as you can then lay aside all the wood-working tools and have only one sort of work to attend to.

The rudder-frame consists of two rectangular frames joined in such a manner that the two planes, or rectangles, cross at right angles.

The horizontal rudder-frame is first constructed by using the six rudder-frame pieces already mentioned, the six-foot lengths being joined by the four two-foot lengths, thus forming a rectangular frame six feet long and two feet wide, with cross-pieces spaced so that they are 2–1/2 feet from each end and one foot apart in the center (Fig. 3).

The corners of this frame are fastened together by half-and-half lap-joints glued and nailed and reinforced by sheet-metal angle-irons (Fig. 3, A).

The vertical rudder-frame is built from the pieces 8 feet 11 inches long for top and bottom beams, connected by the 3-foot-10-inch pieces. One of the latter connects the two long pieces at the extreme ends, with the second two feet from it, and between these, exactly half-way from top and bottom, is fastened the horizontal frame by means of 3/16″ bolts, which can be easily removed. The two frames, when in position, will appear as in Fig. 3, B, and each joint should be carefully made and reinforced with plates of metal or angle-irons, for the rudder bears considerable strain while in flight.

At the points marked BB, BB, eyebolts should be placed through the corners of the beams to receive the stay-wires which support the rudder-frame and stiffen it.

The frame is merely slipped into the sockets, already provided on the cross-pieces of the planes, and is held in position by a bolt passed through these sockets, and to steady it and keep it true the truss-wires are essential. The rudder-frame may be placed in position and afterward covered, or it may be covered first and placed on the frame when finished, which is by far the easier method.

The method of trussing the rudder is simple, and is clearly shown in Figs. 5 and 6:

Two wires run from the top of the vertical rudder-plane to the lower sockets on the rear beams of the plane 4–1/2 feet from each wing-tip (A, A) and two similar wires run from the lower corners of the rudder to the top sockets of the plane of the same stanchions (C). Four other wires (E, F, G, H) brace the horizontal plane of the rudder by extending from its corners to the sockets which support the inner ends of the rudder-beams.

Eight other wires (I, I, in the cut) merely serve to brace the horizontal and vertical planes and hold them rigidly together. These pass from the eyebolts in the corners of the rudder-frames to the eyebolts in the opposite rudder-frame, as shown in Fig. 4, A, A.

### Covering the Planes

Upon the covering you select for the planes and rudders a great deal depends.

Plain, strong silk, unbleached cotton or muslin, plain sheeting, or, in fact, any strong, light, closely woven cloth will serve; but all such materials are liable to tear or rip and allow a great deal of air to pass through their surface.

On the other hand, the various fabrics made and sold for covering gliders and aeroplanes are air and water proof, very tough and strong, and exceedingly light.

They cost only a little more than good silk or linen, and are so much safer and better that it does not pay to use any other material, unless it is impossible to procure the proper fabric.

Nowadays no boy need go without the proper materials or appliances for building a glider or aeroplane, for the industry has become so common that every large city has firms that make a specialty of aeronautical supplies.

Even if you live in the country or far from a large town you may secure anything you wish by consulting one of the aeronautical magazines and writing to advertisers for illustrated lists.

Most of the aeronautical supplies you will require are so light and small that they can be shipped by parcels post and it will surprise you to find how cheaply all such supplies are sold.

In covering the planes some builders cut the cloth into seven strips, each 4′ 6–1/2″ long, and sew these together so that a piece a little over 20 feet long is formed. To reinforce the seams narrow strips 1–1/2″ wide are sewed across the cloth and turned over to form a stout double thickness over each rib.

While this method of construction is very strong, yet the difficulty in getting the seams even and the cloth free from wrinkles is very great, and is apt to discourage an amateur.

A method which will serve every purpose is to cut the cloth into strips a little over four feet in length and glue the end of each strip around the front horizontal beams of each plane, and tack it in place with small copper tacks.

Draw the other end back over the ribs and tack the edges of the strip to the ribs as you go along.

In placing the tacks through the fabric a strip of narrow felt or tape should be placed under the tacks to prevent their tearing out; but this should be almost as narrow as the width of the tack-heads to prevent loose edges, and care should be taken to keep each piece of fabric very tight, smooth, and free from wrinkles as you tack it in position. As the ribs are spaced one foot apart and the cloth is one yard wide, the edges of the cloth should lap on the ribs, and even if you use 24-inch cloth the edges will still lap.

The rear edges of the fabric will, of course, be loose and flexible between each rib, and this is the proper way to have it.

Planes covered in this way are known as "single-surfaced" planes, and a great deal better results are obtained with aeroplanes by covering both upper and lower surfaces of the planes; but for gliders this is not of any importance, and only adds to the weight.

Probably you will find it easier to take the planes apart for covering them; but you are less liable to strain or break the ribs and beams if the covering is put on after the frame is trussed up with wires.

The surfaces of the two rudder-frames must also be covered with fabric, which is stretched over both sides of each frame, passing completely around one end, and is then tacked along all the edges. The last edge should be turned under before tacking in place to prevent any possibility of tearing out.

When your planes and rudders are covered the glider is complete, and all that is necessary is to take it to the proper spot and try a glide or flight.

If the place is comparatively near home you can carry the glider entire; but if at any considerable distance, the rudder should be disconnected and carried separately.

To do this merely remove the bolts that pass through the rudder-beam sockets and loosen the guy-wires connecting the rudder with the machine by loosening the turnbuckles. The latter may then be unhooked and the entire rudder slipped away from the glider.

If you have to transport the glider for a great distance you may take it completely apart in a few minutes, for it is designed to disassemble readily.

### To Disassemble the Glider

Having removed the rudder-frames and loosened all the turnbuckles in the plane sections, unhook the turnbuckles from the eyebolts.

Lift the upper plane from the stanchions and lay it on a flat surface; lift the stanchions from the sockets and tie them neatly together.

Carefully coil up the truss-wires, being careful not to kink or bend them, and tie each coil to its proper socket.

Place one plane on top of the other, taking care that each socket lies against its fellow below, and tie the two planes tightly together.

To prevent any danger of the planes chafing or injuring the fabric, you should have eight short plugs of the same size and shape as the stanchions, and should place one of these in each corner socket and in the four central sockets, thus keeping the planes a short distance apart. The coils of wire may be slipped between the planes, and the turnbuckles tied to the sockets to prevent their shaking about.

If turnbuckles with two hook-ends are used they may be removed entirely and packed separately.

The vertical rudder and horizontal rudder-frames should also be separated and packed flat together.

In this way the entire glider may be packed in a space 20 feet long, 4–1/2 feet wide, and less than 10 inches thick.

It is very easy to set it up again, for everything fits in place, and after a little practice you will know just how tight the wires should be, and will learn to assemble your glider in a wonderfully short space of time.

## Chapter X

### MONOPLANE GLIDERS

AFTER you have built a biplane glider such as has been described, and have learned to use it with confidence, you may try your hand at constructing gliders of other types.

Most of the biplane gliders are so similar to the Chanute type already described that the same general directions for building and using them would serve equally well for any one, the only difference being in details of proportions, shapes of planes, etc.

A very different class of gliders is that of the monoplane type, the most successful and safest of which is undoubtedly the Montgomery glider.

In a way this is not a true monoplane, but a double or tandem monoplane, and it is designed upon such excellent lines and scientific principles that with strong and careful construction and a little care it will prove just as safe as the biplane gliders.

The Montgomery glider is a patented machine, and therefore you cannot reproduce or use it except for experimental purposes without securing permission; but you can alter the design slightly or otherwise change the machine and possibly secure even better results than with the original design.

Very few owners of patents will object to boys making or

using patented machines for themselves, and will usually grant permission for so doing.

The general construction and the detailed measurements are well and clearly shown in Figs. 1, 2, 3. The machine is exceedingly simple; but, as the planes are of the monoplane style, great care must be taken to have all the parts strong and carefully built, as there is no trussing and bracing possible as in the case of biplanes.

The framework consists mainly of two light bars (O, O) tapered off into spars (I, I) and with a heavier lower beam (N) connected to the bars (O, O) by the four diagonal stanchions (H, H).

The wings are built up of frames consisting of two spruce horizontal beams attached by metal clips to the bars (O, O) and having fifty-eight ribs each. The ribs are curved as described for the Chanute glider, and are equally spaced, save for a space of 18 inches in the center of each plane. The covering of the wings is a light rubberized silk made with pockets through which the ribs are passed.

The ribs should be of spruce 1/4" wide x 5/16" deep, and, as the curve is under considerable strain, they are best made from two pieces curved and glued together in a form under pressure.

The wing-bars, or beams, are of hickory 1–1/8" x 1–3/4" at the center, and are tapered to about half this size at the tips.

The frame-bars (O, O) should be of spruce 1–3/4" x 2" at the center, and tapered off to their ends, the forward end being almost pointed and quite a little smaller than the rear end.

The lower bar of the frame (N) is 1–1/4" thick, and from 3" to 3–1/2" deep at the center. The upper edge is straight; but the lower edge is arched, being thickest a little nearer the rear than the front.

Extending from the bar (N) up beside the forward wing-bars, and projecting 19 inches above the bars, are two masts secured by sockets to N, and by lashing and metal clips to the wing-beams. These masts support No. 12 piano-wire stays (F) which support the forward beams of each plane. From the lower side of this forward bar of the wings other wires (F, F) are brought down to N, and these must be set up until the wing-bars are "drooped" or arched as shown in the cuts.

The rear bars of each wing are not trussed, but are hinged at (Q, Q) so that they droop or hang loosely at their ends quite a bit lower than the forward bars. From the ends of these loose rear bars, control-cords (E, E) run over pulleys on N, and are attached to the short stirrup-bar (M), which is 14 inches long and serves as a foot-rest for the operator, who sits on the seat (P), which is merely a padded bar or saddle, and may be made from an old bicycle seat.

By pressing on the right or left side of the stirrup-bar the cords (which are crossed) pull down the opposite wing-tips.

This also aids in balancing the machine; but equilibrium is mainly maintained by the automatic action of the huge tail or fin (C) attached to the rudder (D).

By pressing down on the stirrup-bar equally with both feet all the rear wing-tips are drawn down with a sort of brake effect which causes the glider to land as lightly as a feather.

In addition to these controls a very effective longitudinal control can be obtained by pulling down the cord and pulleys on the rear wing (B) which thus alters the curve of the rear plane as related to the forward plane. The tail, or rudder (D), and the fin (C) are framed from light spruce half-and-half jointed and clamped with aluminum angle-irons and stayed by wires from one edge to the other.

The horizontal tail (D) is set very close to the rear edge of the plane (B), and is hinged to the extremities of O, O. Control-cords (J, K) pass from the center of the rim of this rudder and lead over pulleys at the junction of N and H, and at the top of the mast (G), and hence around a pulley on O, in front of H, and are both fastened to a short wooden bar or toggle (L), which is located on a stationary wire (K2).

Owing to the angular pull of these cords on the opposite ends of L, the latter stays locked in any position that it may be placed on K2. By removing this toggle up or down the wire (K2) the rudder is raised or lowered, and the glider descends or rises accordingly.

This is a very light machine, weighing complete about 40 pounds, and, as the control is mechanical, it is far more like a real aeroplane than the biplane gliders, in which balancing and elevating are accomplished by swaying the body and thus changing the center of gravity.

The large vertical fin (C) moves up and down with the horizontal rudder, but does not move sideways, and acts only as a stabilizer.

On several occasions operators of this type of glider have deliberately turned side-somersaults with it, and have descended from heights as great as four thousand feet, the speed being as high as sixty-eight miles an hour on these glides.

Its equilibrium is almost positive, and if released by itself when at a height and upside down, it will automatically right itself and will glide to earth after falling only four or five times its own spread.

### Using a Monoplane Glider

The biplane glider is very easy to use, and is practically perfectly safe, but in the case of the Montgomery type greater care and caution should be used.

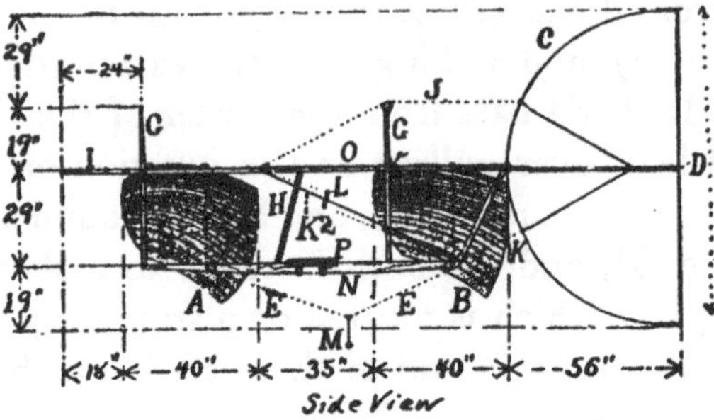


Fig. 1

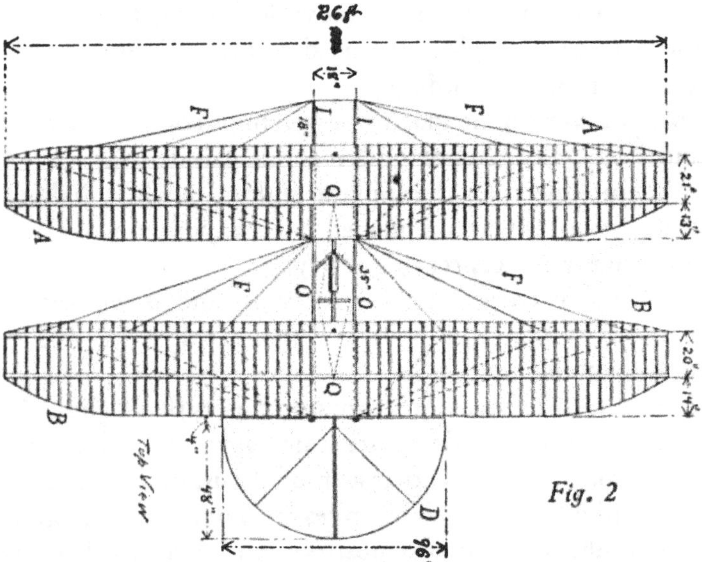


Fig. 2

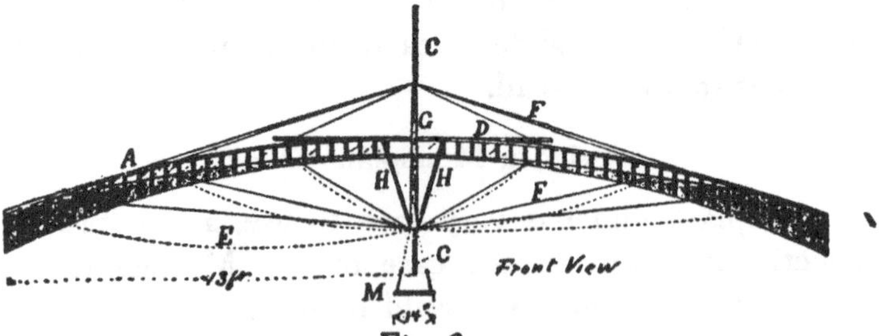


Fig. 3

DETAILS OF MONTGOMERY GLIDER CONSTRUCTION

In the biplane glider the operator is free to use his feet in landing, whereas in the monoplane glider the operator is seated, and the feet are used in controlling the machine, and must be removed from the stirrup-bar and dropped below when landing.

Moreover, to use this machine with the best results it must be dropped from a height, although by having two friends hold it and run with it down a slope very fine glides may be obtained.

Never attempt flights from any great height, and if possible practise over water, where a fall will not injure you.

Even with the best of care accidents will at times occur, especially when learning, and it is wise to be cautious and make haste slowly.

### Model Gliders

Either biplane or monoplane gliders may be made in small model sizes, and, while miniature reproductions of man-carrying gliders will almost always work excellently, yet gliders designed as models and of two or three feet spread will not always work at all well when carried out in full size.

This is as true of gliders as of model aeroplanes or fliers, and, while a well-recognized fact, the explanation is so technical and depends so largely upon mathematical problems that it would be very difficult to make it clear in simple, understandable language.

Boys may make very interesting experiments with model gliders, however, and if one is designed which possesses extraordinary lifting and gliding properties it would be well worth making a full-sized machine of the same type and trying it out.

This is the only way that you can be absolutely certain

that your design will or will not prove efficient as a man-carrier.

Any boy who has designed or built model fliers can readily build a model glider, for the methods of construction are identical, practically the only differences being in the shape and size of planes and the omission of propellers and rubber-band motors.

The materials, wood and fabric, used are the same for the model gliders as for model aeroplanes, and to fly them one merely attaches a suitable weight to the lower part of the machine and drops it from a height facing the wind. In a way the flight of a model glider is even more interesting than that of a model flier, for it sails so gently and easily without any propulsive force of its own that it seems almost endowed with life.

# Part IV

# THE MODERN AEROPLANE

# Chapter XI

## TYPES OF AEROPLANES

THE true aeroplane is practically nothing but a strongly built and perfected glider provided with a motor and propeller to drive it through the air. There are many points of difference between the simple glider and the latest types of aeroplanes, however, but these points are nearly always details which are made necessary owing to the greater weight to be carried and the more perfect control required in the aeroplane.

The Wright brothers developed their biplanes from a biplane glider very similar to the Chanute type, already described; and, in fact, the earlier Wright machines were merely biplane gliders with a motor and propellers.

Even the light and simple Montgomery glider has been equipped with motor and propeller, and has made excellent flights.

Aeroplanes vary in size from the tiny *Demoiselle* of Santos Dumont, which spreads but 16 feet and weighs 248 pounds, to the huge 50-foot passenger-carrying aeroplanes built recently.

Many of the most successful and fastest aeroplanes are far smaller than the aeroplanes one commonly sees, for these machines are very high-powered, and if the power of an aeroplane is increased the spread or surface of the planes may be decreased.

The faster an aeroplane travels the more a given area will sustain, and in building racing-machines power has been carried to extremes, and wing area reduced until it seems impossible that the machines can actually fly at all. Nevertheless, Vedrines won the Gordon Bennett prize at Chicago in a machine spreading but 18 feet, and traveled at the amazing speed of 109 miles an hour.

The power that drove this aeroplane was a 140-horsepower motor, however, whereas the power required to safely lift and fly an ordinary machine with a spread of 35 feet or more need not be over 35 or 40 horse-power.

So if you see a very small aeroplane flying swiftly you may be sure it is a high-powered machine, whereas if you see a broad-winged machine traveling slowly but surely—perhaps with several passengers—you may be quite sure that it is a low or medium-powered machine, and very safe and steady.

The Wright machines are of this class, and, in fact, the Wright biplane has developed less and has remained more like the first model than any other type of aeroplane.

Americans are very partial to biplanes, and until recently nearly all the machines made or flown in this country were of this class.

Nowadays, however, a great many excellent monoplanes are built and used in this country, but still the great majority of American aeroplanes are biplanes.

Biplanes are aeroplanes having one plane or wing above the other, as in the Chanute glider. Fig. 1 shows a typical form of biplane.

Monoplanes have but a single wing or plane, like the Bleriot (Fig. 2); but some models have a second plane behind the main plane, just as in the model fliers or the Mont-

*Fig. 1*

*Fig. 2*

gomery glider. These are known as "tandem-surfaced monoplanes."

In addition to monoplanes and biplanes, machines with three or more wings, one above the other, have been built and flown. These are known as "triplanes" and "multiplanes," and while they have been flown successfully, they are seldom used, and have never proved as practical as the single or two winged forms.

## Parts of Aeroplanes

The parts of an aeroplane are mainly the frame, or "chassis"; the body, or "fuselage"; the wings, or "planes"; the rudder and tail; the elevator and the running-gear, or "alighting-gear," in addition to the motor and propeller and the control system.

The chassis, fuselage, planes, etc., are similar in construction and detail to the gliders, but in order to withstand the shock of landing the parts are made much heavier and stronger, and a running-gear is provided.

The running-gear of the original Wright machines consisted merely of wooden runners or skids, and the machine was started on a track or runway.

Nowadays all aeroplanes are provided with wheels as well as skids, and by using large pneumatic tires on the wheels, as well as rubber bands and coiled springs, a great deal of the shock of landing is obviated, and the machine can run over fairly rough ground when starting.

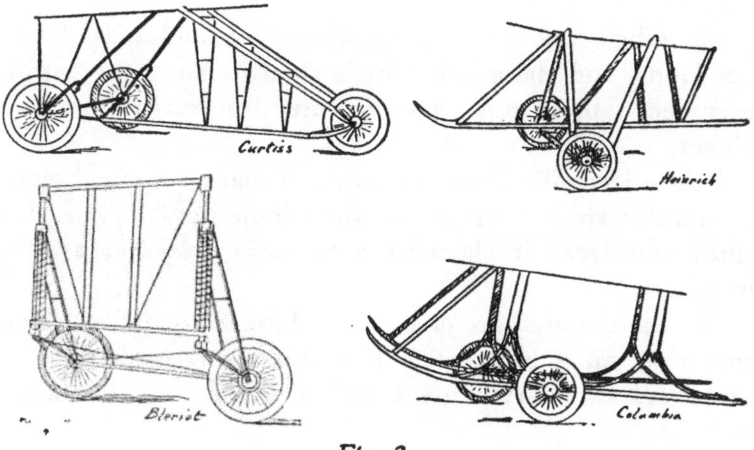


*Fig. 3*

The running-gear varies a great deal in detail in different types of aeroplanes, and each maker has his own ideas of the best form.

Typical forms of running-gear are shown in Fig. 3, and by studying these you can obtain a very good idea of the way the leading aeroplanes are equipped for landing and rising from the ground.

Most aeroplanes are built of wood with metal clamps, sockets, and wire stays, but quite a number have the framework composed mainly of hollow-steel tubing.

Others have the body built of pressed fiber, or are built of thin wood like a boat; but, strange as it may seem, no metal is as strong for its weight as selected spruce, ash, and other woods, and practically all standard machines use wood for their framework, ribs, etc.

Aluminum is used only where light weight without great strength is required, for this metal, as well as its alloys, is not strong enough for most parts of aeroplanes.

Steel sockets and clamps, steel bolts, exceedingly strong steel piano-wire, and steel-wire cables are used wherever possible.

The fabric used for covering wings and rudders is of various kinds, and there is but little choice among the various prepared fabrics now manufactured especially for aeroplanes.

Rubberized silk, linen, cotton, and many patented cloths chemically treated are used, while some builders use plain cloth and treat it themselves to make it water and air proof.

The motors used are of so many different types and kinds that they must be treated in a separate chapter, and the propellers also vary a great deal in shape, size, pitch, and material.

Most aeroplane propellers are constructed of wood, carved out of blocks formed by gluing several thicknesses of different kinds of wood together under pressure.

These are known as "laminated" propellers, and by using soft and hard woods in their construction the outer edges

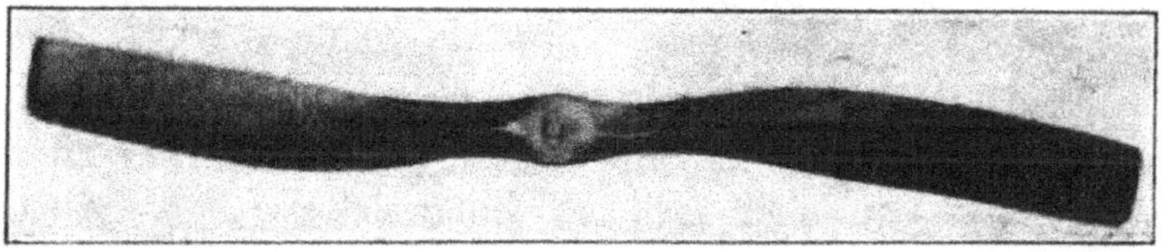

*Fig. 4*

*Fig. 5*

can be made to resist chipping and injury, while the inner portions may be constructed of soft wood, and thus be far lighter than if built entirely of hard, strong wood.

A typical form of propeller is shown in Fig. 4. This is the Charavay type, and has proved very effective in a great many machines.

Most aeroplanes have but one propeller, attached directly to the engine-shaft; but the Wrights, Burgess, and several others have two propellers geared to the engine-shaft with sprockets and chains, and revolving in opposite directions much slower than the motor (Fig. 5). Whether rapidly-revolving single propellers or slowly-revolving twin propellers are the better is a question never yet definitely decided.

The earlier machines and practically all biplanes up to 1912 were equipped with propellers in the rear of the

*Fig. 6*

*Fig. 7*

machine (Chapter III, Fig. 2). But within the past year the tendency has been to transfer the propeller to the forward end of the aeroplane, where it "pulls" the machine through the air instead of "pushing" it, and so in reality it becomes a "tractor" instead of a "propeller."

Nearly all monoplanes have always been built with the propeller in front, and this permitted a covered body or "fuselage" of graceful shape, which made the monoplanes look like huge dragon-flies or some kind of birds.

As soon as biplane builders commenced to place the propellers in front instead of behind the machine they realized that the covered and compact fuselage of the monoplane could be used in a biplane, and one may now see scores of tractor biplanes which are just as graceful and birdlike as the monoplanes, and, in fact, when flying overhead one can scarcely distinguish a biplane with a tractor propeller from a true monoplane (Figs. 6 and 7).

Whether the tractor is really more efficient than the propeller is a matter still under discussion by designers and builders, but it is certain that many propeller biplanes which have been altered to the tractor form have proved far safer, faster, and more reliable when the propeller pulls instead of pushes the machine.

## Chapter XII

### BIPLANES AND MONOPLANES

THE aeroplanes in use to-day are practically all biplanes and monoplanes, and both types prove equally efficient and reliable in the hands of their operators.

There is little real choice between the two kinds of machines; but as a rule the biplane is a steadier, safer, and slower machine than the monoplane, and is far better adapted to beginners or amateurs. The fact that two planes are placed one above the other allows a very simple and strong method of bracing and trussing, as each plane helps to support and strengthen the other.

Monoplanes, on the other hand, must depend for support of the wings upon very strong and rather intricate construction and wire stays fastened to masts, or "pylons."

Biplanes also have the advantage that their lateral balance can be maintained by small auxiliary planes, or ailerons, and consequently stiff and rigid planes may be used, whereas monoplanes usually depend upon flexible or warping wing-tips, which require special methods of construction and mechanical devices.

In order to lift and fly with a certain weight and a given power a definite area of wing surface must be provided; hence the monoplane requires longer planes than the biplane of equal plane surface, which has the area of surface divided into two sections, and this again necessitates a

*Fig. 1*

THE WRIGHT BROTHERS

more elaborate and complicated system of stays and braces.

For these reasons biplanes are and always will be far easier to build than monoplanes, and, moreover, they are less liable to break or to develop weak spots than the single-winged machines.

### Leading Biplane Types

Probably the best-known biplane in the world is the Wright. This is not because the Wright machine is any better or more widely used than several other types, but merely because it was the first really successful aeroplane, and the one which opened the eyes of the world to the fact that men had learned to fly.

To the Wright brothers, Orville and Wilbur (Fig. 1), must be given every credit and honor for having been the pioneers of aviation; and yet to-day the Wright machine is by no means as efficient or fast as many other aeroplanes.

The Wright machine is a rather heavy, slow, and cumbersome affair, with many features unchanged since the earliest models; but it is a favorite model with many exhibition fliers, and especially with aviators who make a specialty of carrying passengers.

The Wright machines are made in several models, but all are very similar in appearance and are readily distinguished from other biplanes by the rounded wing-tips, boxlike vertical rudders, and upturned skids to the running-gear, as well as by the twin propellers and lever-controls (Fig. 2).

The Burgess-Wright (Chapter III, Fig. 2) is a modified Wright machine preferred to the latter by many operators. This machine is used by the United States army and navy, and by many long-distance and cross-country fliers. Its

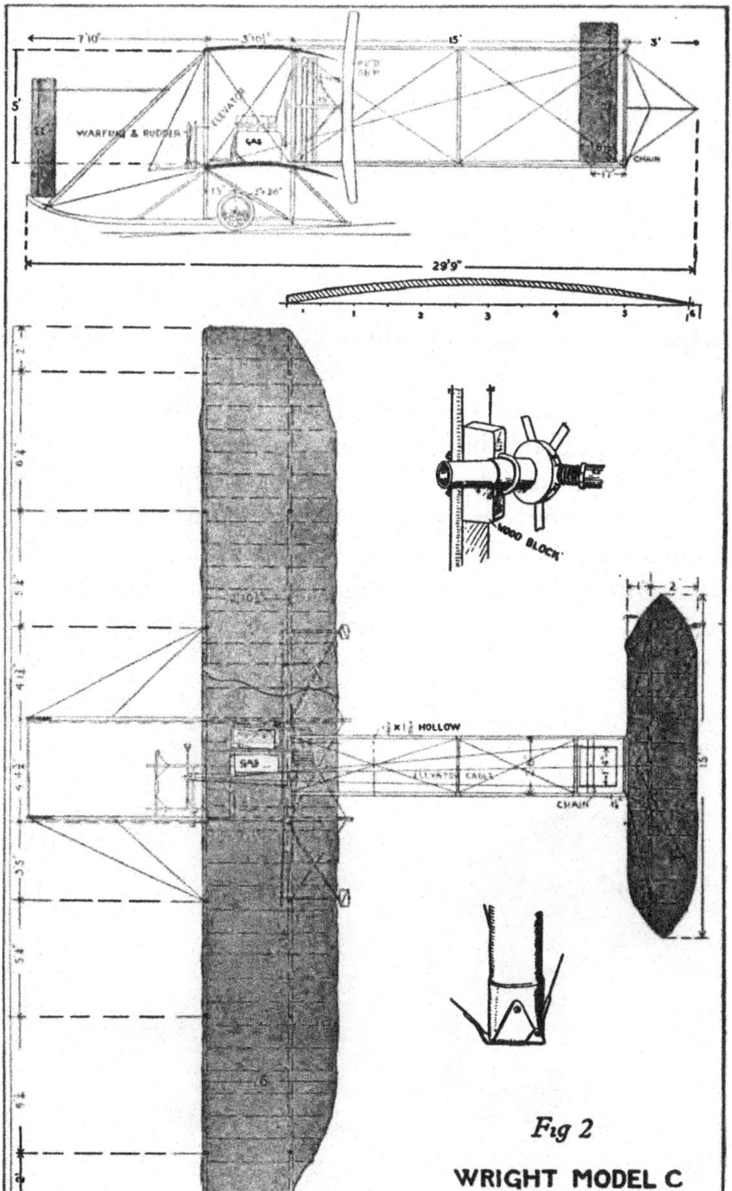


*Fig 2*

**WRIGHT MODEL C**

most characteristic features are the two triangular vertical planes in front of the lower plane—a feature also found in many genuine Wright machines.

Burgess machines are, however, now made in the tractor form, and these are scarcely recognizable as Burgess aeroplanes.

Almost as well known as the Wright aeroplanes are the Curtiss machines, made world-famous by their record flights in the hands of their inventor, Mr. Glenn Curtiss, and that most spectacular of aviators, Mr. Lincoln Beachy.

It was a Curtiss biplane that first flew from Albany to New York, from New York to Philadelphia, from Key West to Havana, and that first flew from and landed on the deck of a warship, and it was also the first American hydroaeroplane to rise successfully from the water.

It was in a Curtiss machine that Beachy made his famous

*Fig. 3*

flight through the gorge at Niagara, and the Curtiss machine is renowned for speed, safety, and ease of handling.

Instead of the flexible warping wings of the various Wright types the Curtiss uses ailerons between the tips of the wings, and in place of the clumsy box-tail a graceful "pigeon-tail" or kitelike tail is used (Fig. 3). The shape of the running-gear, the ailerons, and the control easily distinguish the Curtiss machines even when built in tractor form (Fig. 4). The latest Curtiss models do not have the "head" or front elevator, shown in Fig. 3, which was used on all the earlier Wright and Curtiss machines, but which has now been generally discarded.

In addition to these two widely used types of biplanes there are many others less familiar to the average person, but of splendid efficiency and powers of flight.

Among these are the Benoist (Fig. 5), the Thomas, the Baldwin, the Witteman, the Kirkham, and dozens of others, many of which hold world's records and are superior in construction and design to many better-known types.

A great many are, however, merely Wright or Curtiss types slightly altered, or with additional or unique features designed to improve them. Many of these machines are known by a special name, but to all intents and purposes they are Curtiss or Wright machines.

Another very distinct type of biplane is the Farman (Fig. 6). In this machine the ailerons are flaps of fabric hinged to the rear edges of the planes, while the tail consists of two miniature planes like a small biplane, between which are the vertical rudders. Small flaps at the rear of the two horizontal tail-planes act as horizontal rudders and a broad plane in front acts as the elevator, or head.

Many biplanes which have flown splendidly are really combinations of the Wright, Curtiss, and Farman types.

Fig. 4

FIRST CURTISS TRACTOR

Such, for example, is the famous Baldwin *Red Devil*, flown by Cecil Peoli, the "Boy Aviator."

This machine has the Farman type of ailerons, the Curtiss control, with some details of the Wright, while many new ideas and original improvements have been added, such as steel-tube construction, peculiar rudders and tail, etc.

A form of biplane which is very distinct from all others and is absolutely unique in many ways is the Boland Tailless

*Fig. 5*

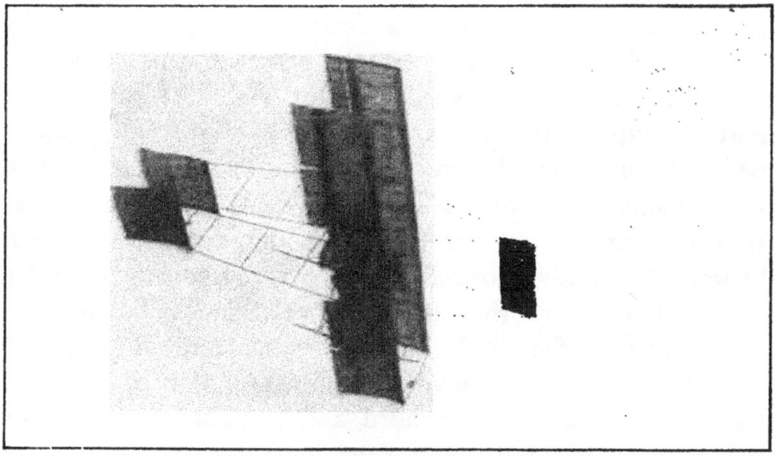



*Fig. 6*

Biplane, invented by the late Frank Boland, and improved and perfected by the Boland Aeroplane and Motor Company (Fig. 7).

When first flown by its inventor the machine absolutely amazed all beholders, for it performed evolutions and flew in a manner apparently contrary to all laws of aviation, and its intrepid owner drove the odd creation with the same facility and ease with which he would handle an automobile.

The machine was flown successfully throughout several months, and was exhibited in South America and the West Indies, and in the island of Trinidad the unfortunate inventor lost his life.

The machine is so unusual and so distinct from all others that a detailed description and plans are of value to all interested in aeroplanes.

The principal and most striking feature of the machine is the fact that it is "tailless," and, although various other machines of the "tailless" form have been constructed, yet none have operated in the same manner as the Boland.

The "nacelle," or body, contains the engine, pilot, and passenger, and consists of a boatlike structure near the center of the machine between the planes. This body is of oval section at the forward end but V-shaped at the rear seat and tapers off to a sharp edge at the rear. The engine-bed rails run from the cross-piece which forms the back of the passenger seat to a point at the rear, where they meet the tapering sides of the body just above the lower wing-beam. The body frame is metal, covered and upholstered, and within it are the steering-wheels, gages, tachometer, etc. Thirteen feet and eight inches in front of the main planes the elevator is pivoted. This is a curved plane 12 feet wide with 3-1/2-foot chord, and in flight it carries fully half the weight of the entire machine, but its location far ahead of the body gives it so much leverage that the actual weight supported is but 120 pounds.

By rocking the yoke of the elevator forward or backward the machine is made to rise or descend.

The most noteworthy feature of the machine is the method of control. Instead of ailerons, warping-wings, or rudders, the Boland biplane has two "jibs," one at each end of the planes, pivoted on an oblique axis and normally in a vertical position. Each of these jibs works in but one direction—inward—and is controlled by a 5/32" cable leading around the steering-wheel. By moving this wheel, exactly as in steering an automobile, the machine is turned to right or left, and as the action of the jib tends to retard the machine on the side which is pulled, the jibs act as lateral stabilizers as well as rudders.

By turning the wheel to the right the machine turns to the right and tips or "banks" just enough to prevent side-slipping or capsizing. Any tendency of the biplane to tip or bank when traveling through the air may be corrected by pulling in the jib on the high side. If this is not done, however, the machine merely turns a little from its course.

The running-gear consists of two skids extending far out in front and serving as a support for the elevating-plane. There are two 20" x 3" wheels on a long axle situated 18 inches in front of the forward edge of the main planes, and rubber shock-absorbers are also provided.

The power-plant is a Boland "V" motor of 60 horse-power driving a single propeller.

The main planes spread 35.5 feet and are 5.5 feet wide and 5.5 feet apart.

The weight of the machine complete with gas, water, and oil is about 900 pounds, and it is capable of traveling at about 60 miles an hour.

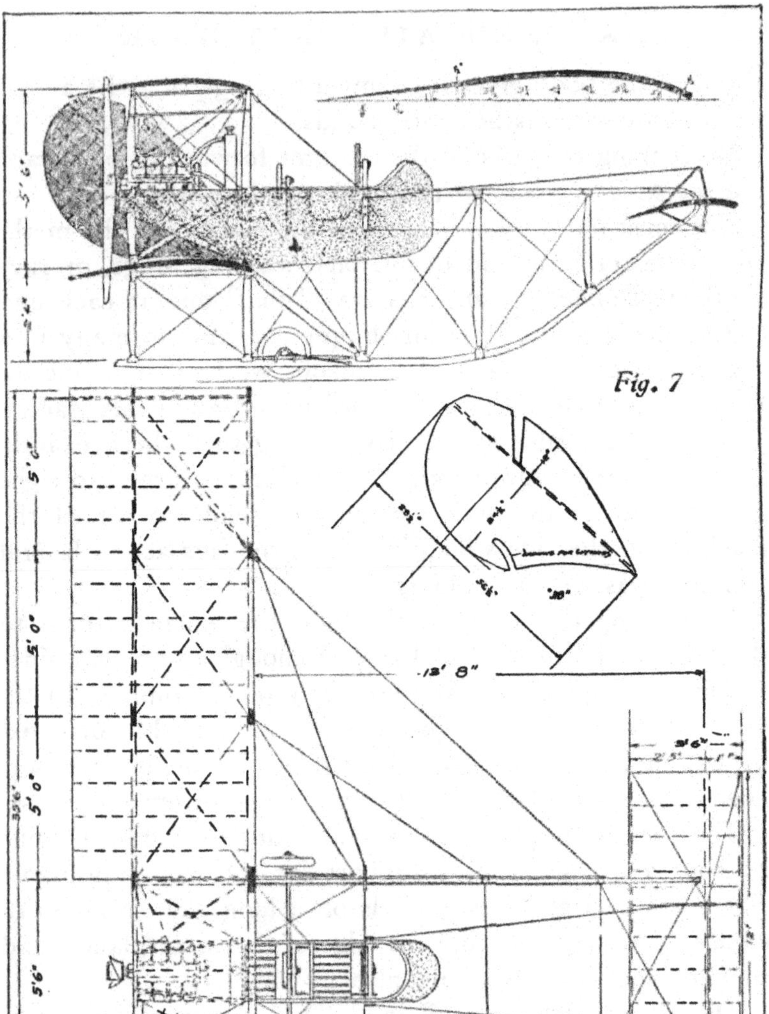

Fig. 7
BOLAND TAILLESS BIPLANE
13' 8"
5' 6"
5' 0"
5' 0"
5' 6"
2' 0"
5' 6"
3' 6"
1612

## Typical Leading Monoplanes

Various as are the different biplanes in use, even more numerous are the designs of monoplanes.

Most of the monoplanes are of French origin, and as the first to prove successful was the Bleriot, so to-day the Bleriot type is by far the most familiar and most widely known form.

The Bleriot (Fig. 8) is a very graceful and efficient ma-

*Fig. 8*

*Fig. 9*

*Fig. 10*

*Fig. 11*

*Fig. 12*

HEINRICH MONOPLANE

chine, and, although minor details of construction, controls, tails, etc., have been altered from the original form, yet it is always readily identified by round-tipped wings, skeleton body, square, vertical rudder, and peculiar landing-gear.

Many excellent Bleriot-type machines are now built in this country, and to the layman these are indistinguishable from the imported Bleriots.

Another monoplane of French origin which is almost as renowned as the Bleriot is the Antoinette (shown in flight in Fig. 9), although this type is less often seen now than in the earlier days of aviation.

The Antoinette is a splendidly efficient machine, and cannot be mistaken for any other design, as the square wing-tips (ending in a scroll), the boatlike body, peculiar tail, with the radiators beside the body and beneath the wings, give it a unique and striking appearance.

Caudrans, Deperdussins, Moranes, Grades, Nieuports, and dozens of other European monoplanes are in daily use in America, and each differs from all others in some detail of construction or design, even though it may be very hard to distinguish them at first sight.

American monoplanes of several makes have now taken their place with those of French design, and the Heinrich, Moissant, Columbia, and others, are just as good fliers and just as well built as the European machines. The Heinrich (Figs. 10, 11, 12, 13) is a splendid machine, very distinct in design and appearance from most other American monoplanes, which as a rule are modifications of French types.

The Heinrich has a completely inclosed fuselage, with a running-gear which combines the good points of both the Bleriot and the Antoinette systems. The wings are semi-rounded at the tips, and the body is very narrow, thus re-

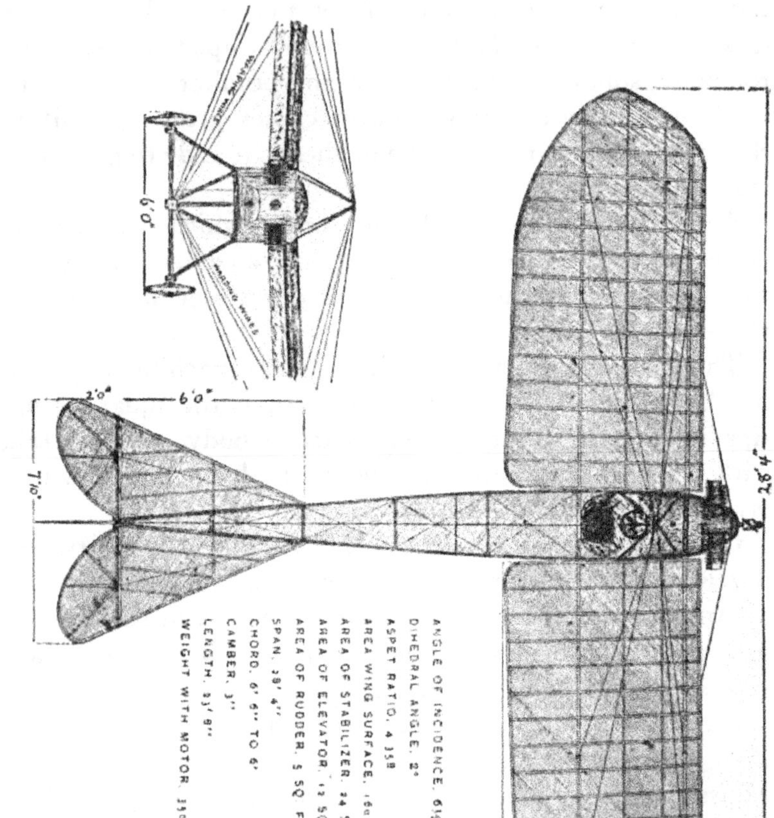


***Fig. 13***

**HEINRICH MONOPLANE MODEL D**

ducing head resistance to the minimum. The tail is broad, rounded, forked, and birdlike, with hinged horizontal rudders at the rear and a small vertical rudder between them. The whole machine is very graceful in form, and has proved a most reliable and efficient flier.

Many designers have departed radically from established principles and have produced monoplanes of odd or freakish form. Among these are the Gallaudet "bullet" (Fig. 14) and the Waldron monoplane. The "bullet" was

*Fig. 14*

a fish-shaped machine of steel-tube construction, and with the three-bladed propeller in the rear. It was designed to produce great speed, but was not a success.

The Waldron monoplane, on the other hand, was a machine which had ailerons like a Farman biplane, with a running-gear and control similar to the Curtiss, and with the propeller behind the wings. It was in effect a biplane with only the upper plane, and, while it made a number of successful flights, it was always a "cranky" and dangerous machine. Such "freaks" are constantly appearing, and they all serve an excellent purpose in demonstrating just what is practical and what not in aeroplane design. Only in this way can better results be accomplished and progress made in aviation, for the best theories on paper often prove failures in practice.

Of very "freakish" appearance but of wonderful power and efficiency are the specially designed racing-monoplanes. The Deperdussin racer, in which Vedrines won the Gordon Bennett cup at Chicago, is a machine of this type (Fig. 15).

This monoplane is constructed with a fish-shaped body built of thin basswood strips 1/8″ thick, glued together in alternate layers over a form which is removed after the body is complete. The wings are very short and small, and spread but 18 feet, while the motor is a "Gnome" developing 140 horse-power. The running-gear is very simple, and even the wheels are covered with disks of aluminum to re-

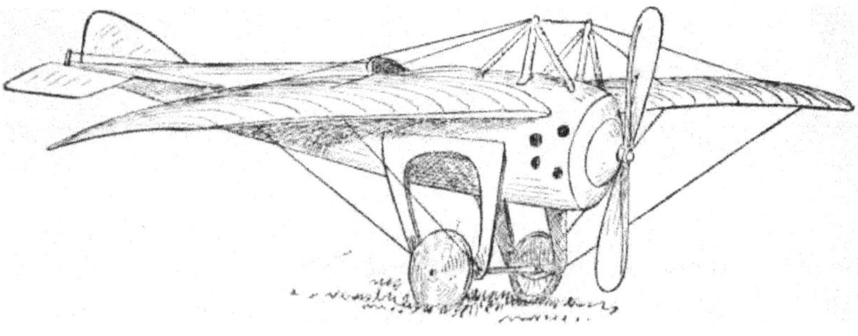

*Fig. 15*

duce air resistance. With this machine the fearless French aviator dashed through the air at 109 miles an hour, looking, as one observer stated, "like a huge, winged cannon traveling breech first."

Other machines of similar construction have since been

built, and, while all of the type are excessively fast, yet they require the utmost skill in handling, and are most dangerous machines in the hands of a novice.

Their only value is for racing and reaching high altitudes, and for general utility they are not half as good as the prosaic and slow-going Wright or Curtiss biplanes or the reliable Bleriot or Heinrich monoplanes.

## Chapter XIII

### THE HEART OF THE AEROPLANE

THE heart, lungs, soul, and life of any aeroplane are, so to speak, the light, compact, and powerful gasolene-motor. Many people seem to think that the motor is a secondary consideration, and that, having built or bought a plane, they can place an old marine or automobile engine in it and fly. Many boys and men have done this, and because their machines proved dismal failures they have become discouraged and disgusted.

As a matter of fact, the motor for an aeroplane must possess many features and details that are very distinct from those of either marine, stationary, or automobile engines. Whereas the automobile engine is constantly varying in its work and seldom operates at its maximum power or speed, the aeroplane motor works at a uniform and very high speed, and is called on for its maximum power all the time it is in operation. To withstand this strain the strongest and most wear-resisting materials must be used in its construction and unfailing methods of lubrication and cooling are absolutely necessary.

As an aeroplane motor must be exceedingly light and free from vibration, you can readily understand how difficult it is to combine these various requirements. It is very easy to construct a reliable engine fairly free from vibration and with long life, provided it weighs from eight to fifteen pounds or more per horse-power, and few automobile motors weigh less than this. Imagine trying to build a motor of the same power, reliability, and life and weighing but 2–1/2 to five pounds per horse-power; or, as a better illustration, think of a six-cylindered, 60-horse-power motor so light in weight that a man can readily lift it in his hands!

Yet this is actually accomplished in many aeroplane engines; and the wonder is, not that aviation motors are unreliable, but that they are half as dependable as they are.

In place of the cast-iron, heavy forgings, stout, long bearings, heavy supports, and thick cylinder-walls of the marine and automobile motor, the aeroplane engine is constructed of the highest grade chrome and venadium steels, magnalium metal, manganese bronze, and other selected materials, with delicate ball-bearings, feather-weight crank-cases, and cylinder-walls hardly as thick as ordinary card-board.

All this means a very expensive engine, for no part can be slighted, and the best possible work and utmost care must be devoted to each and every part and detail.

Some idea of the work required to build a modern aeroplane motor may be obtained by considering the crank-shafts and cylinders of any of the best of the aviation engines.

These parts are machined from solid steel bars, and the crank-shaft, in the rough, may weigh as much as 150 or 200 pounds, but when finished weighs less than 25 pounds.

Many air-cooled engines have the cylinders turned and bored from solid steel or iron bars six inches or more in diameter, and yet the completed cylinder weighs but a few ounces, and is less than 1/16″ in thickness.

The work and time required to turn down and finish the many parts of such a motor are what make it expensive; and, whereas a good auto or marine engine may be purchased for $100 to $300, a fairly reliable aviation motor will cost from $1,500 to $6,000, or more than the complete boat or automobile, engine and all!

There are a great many types, models, and forms of aeroplane motors on the market, some good, some fair, and some worthless, and they are made both in America and in Europe.

Many Europeans use American motors, and a great many American aviators use French motors; but, as a matter of fact, American motors as a whole have not been as highly perfected as those of European manufacture.

Aeroplane motors are of both two and four cycle types, and before describing them further it may be well to explain just what the difference is between a four-cycle and a two-cycle motor, and the action and principle of each.

### Operation of Two-Cycle Motor

The two-cycle or two-stroke motor is a form of engine in which the piston receives an impulse or explosion *once in each revolution of the crank, or once on every two strokes of the piston and connecting-rod.*

It is a very simple, easily understood form, with few parts, and by referring to Figs. 1, 2, 3, 4, you can readily understand just how the motor operates.

Fig. 1 represents the motor in section, with the piston (A) at the bottom or end of its stroke. As the crank revolves and the piston (A) travels upwa d in the cylinder (B) a charge of gasolene and air is sucked in from the carburetor through the opening (C) and fills the crank-case (D), from which it cannot escape, as the only opening is the port (E) which is covered by the piston, as shown in Fig. 2.

The crank continuing to revolve, the piston (A) again de-

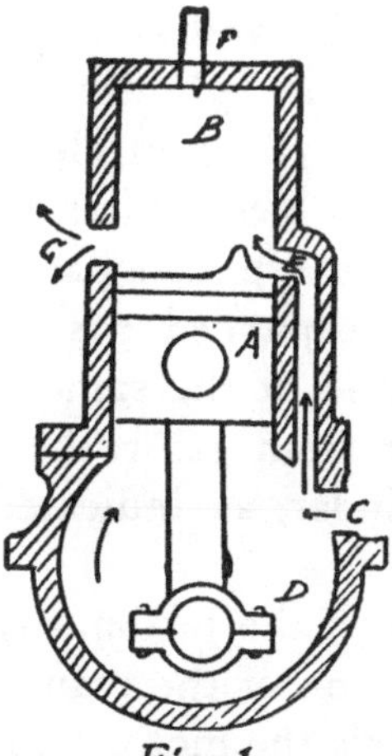


*Fig. 1*

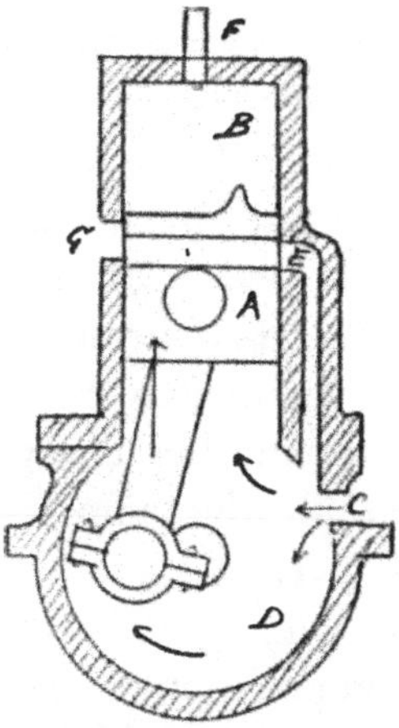


*Fig. 2*

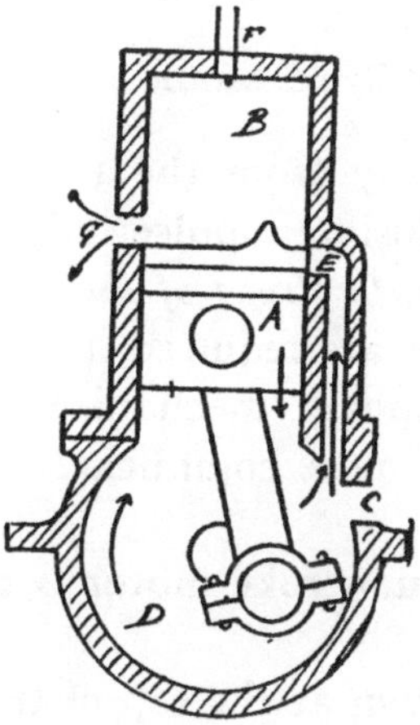


*Fig. 3*

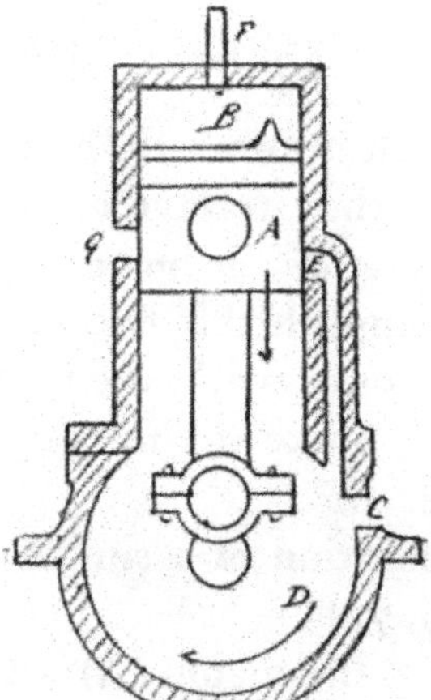


*Fig. 4*

scends in the cylinder and compresses the charge of gas in the base, and at the point shown in Fig. 3, the port (E) is uncovered and the compressed gas rushes up and into the cylinder, as shown by the arrows.

The crank-shaft, still revolving, carries the piston on its

upward stroke, closing the port (E) and compressing the charge of gas in the upper part of the cylinder until in the position shown in Fig. 4. At this point an electric spark is produced at the spark-plug (F), which ignites the charge and drives the piston down. As the piston passes the exhaust-port (G) the burning and expanded gas rushes out through the opening to the air, and at the same time, or a moment later, another charge of fresh gas rushes up from the base to fill the empty cylinder, as indicated by the arrows in Fig. 1.

Thus, on every upward stroke of the piston, or, in other words, on every complete revolution of the crank-shaft, an explosion is obtained which propels the motor and develops the power.

### Operation of Four-Cycle Motor

Four-cycle motors differ mainly from the two-cycle inasmuch as they receive an explosive impulse only on *every second revolution, or on every fourth stroke of the piston.*

To accomplish this two valves are required in each cylinder, and these, with their attendant mechanism of gears, cams, etc., make the motor far more complicated than the two-cycle.

The operation of a simple four-stroke motor is shown in Figs. 5, 6, 7, 8.

In Fig. 5 the piston (A) is shown at the top of the stroke, with the inlet-valve (I) beginning to open.

The piston moving downward in the cylinder (B), a charge of fresh gas is drawn into the *top* of the cylinder from the carburetor through I, and fills the cylinder.

At the bottom of this first stroke, as shown in Fig. 6 the inlet-valve (I) closes, and the exhaust-valve (X) also being closed, there is no chance for the charge of gas to escape, and the piston moving up compresses the gas, which is ignited and explodes when the piston reaches the point shown in Fig. 7. The explosion drives down the piston, and at the bottom of the stroke the exhaust-valve (X) opens, and through this valve the burned gases are forced

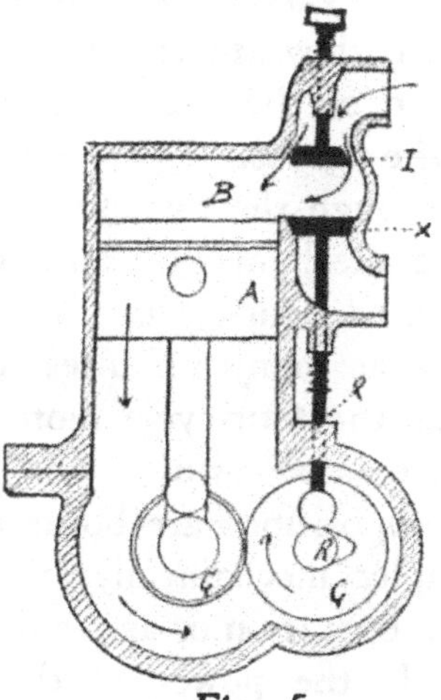


*Fig.* 5

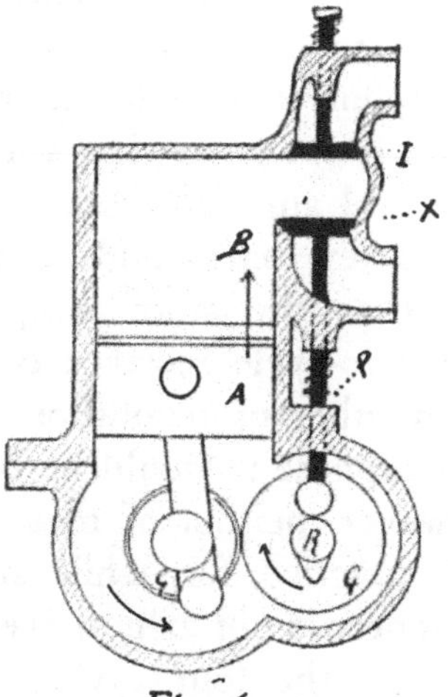


*Fig.* 6

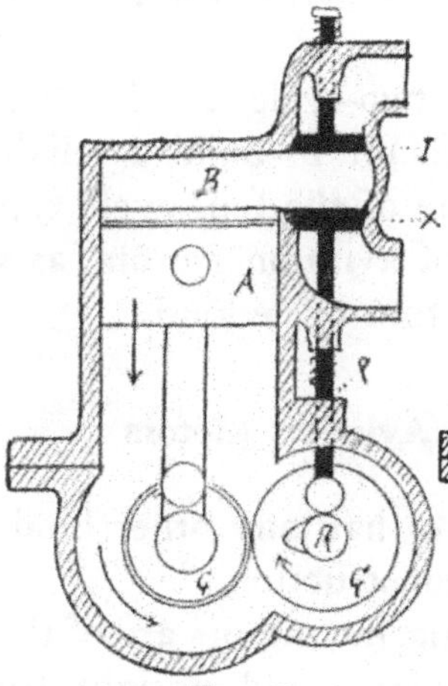


*Fig.* 7

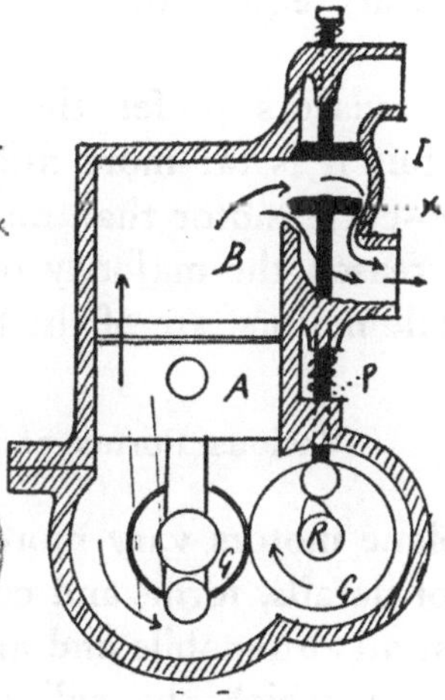


*Fig.* 8

to the open air by the ascending piston, as shown in Fig. 8, until the piston reaches the limit of the stroke, when the exhaust-valve (X) closes and the inlet (I) opens and the operation begins all over again, as shown in Fig. 5.

The inlet and exhaust valves are operated by a cam and rod (R, P) in the figures, and as the valves open but once for each two revolutions of the shaft, they are driven by a gear which revolves at half the speed of the shaft. The gears (G, G) are shown as circles in the figures.

It would appear at first as though the two-cycle engine, with its few parts and simple mechanism, would be far superior to the more complicated four-cycle. Moreover, as the two-cycle receives twice as many impulses for the same number of revolutions as the four-cycle motor, you would suppose it would be twice as powerful.

However, neither of these suppositions are borne out in actual practice, for, while simpler mechanically, the two-cycle motor is not as positive in operation or as readily controlled as the four-cycle, while the power of the four-cycle is fully equal to, if not superior to, the two-cycle motor.

Many aviators prefer the two-cycle to the four-cycle motor; but it is far more difficult to build a reliable and good two-cycle motor than one of the four-cycle type, and for that reason the majority of aviation motors, as well as automobile motors, are of the four-cycle kind.

### Various Forms of Aviation Motors

Aeroplane motors vary more than any other kind in the matter of details, form, and construction.

Almost all automobile and marine motors are of the vertical form, in which the cylinders stand upright from the base, one behind the other, although a few are horizontal, with the cylinders opposite.

Aeroplane motors not only are made in the vertical and horizontal forms, but in V-shaped, star, radial, and other

*Fig. 9*

forms, the object of each of these being to save space and weight.

In a vertical motor (Fig. 9) the crank-case must be as long as the total length of the added cylinder diameters. If the cylinders are placed in V-shape (Fig. 10) two piston-rods may be coupled to each crank, and the length of the crank-case greatly reduced, with a consequent saving in weight.

If the motor is of opposed horizontal form (Fig. 11) both the pistons may be attached to the same crank, and the case can be almost as short as for a single-cylinder motor.

In the star form (Fig. 12) and the radial form (Fig. 13) several piston-rods are connected to a single crank, and the case is but little longer than in a single-cylinder motor.

Aside from this reduction in size and weight better balance and more even action can be obtained by arranging the cylinders in the above forms; but the power obtained is

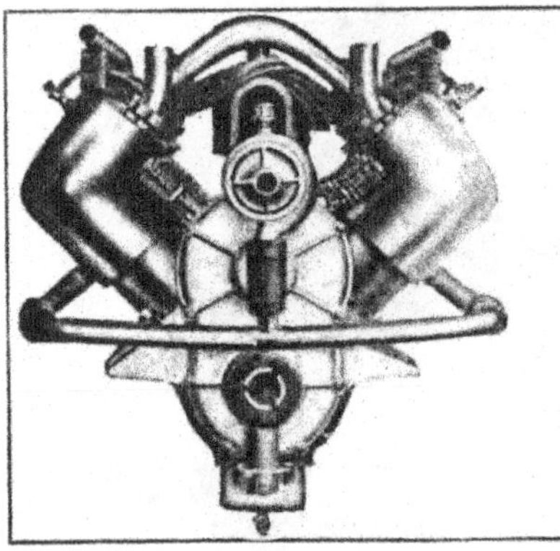

*Fig. 10*

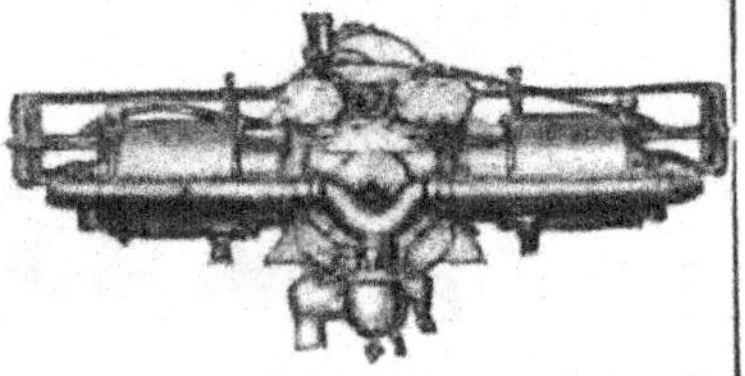

*Fig. 11*

*Fig. 12*

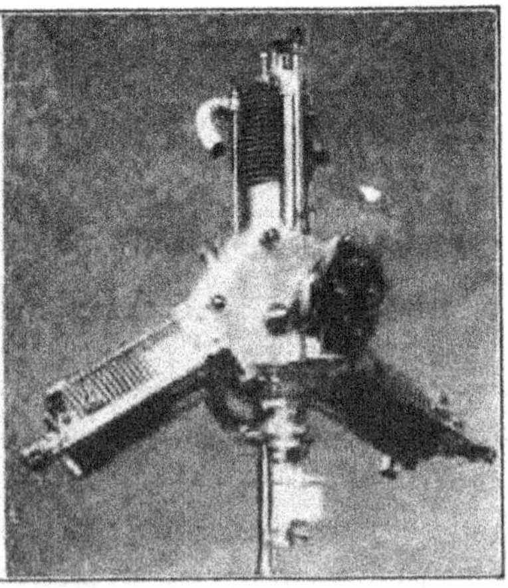

*Fig. 13*

practically the same no matter how the cylinders are placed around the shaft.

All these motors are so-called "stationary," for the crank-case and cylinders are stationary, and only the crank-shaft revolves.

A very different form of motor which has become a great favorite with aviators is the "rotating"-motor.

In this motor the operation is very different from any ordinary motor, and it is well to understand just how it works,

*Fig. 14*

*Fig. 15*

as it is a great puzzle to many people, although really very simple.

### Operation of the Rotating-Motor

There are various kinds of rotating-motors, both American and European, and, while these differ slightly in detail, yet in their operation and principle they are so similar that an explanation of one will serve for all.

Perhaps the best known of all rotating-motors is the

Gnome (Fig. 14); but the Gyro made in this country is very much like it, and in some ways is much superior. A Gyro seven-cylinder motor of 50 horse-power is shown in Fig. 15, with a section of one cylinder in Fig. 16, and a diagrammatic section of the entire motor in Fig. 17.

In this motor the crank (S) is fastened immovably, and the various cylinders are attached to a case which revolves freely around the shaft. Within the cylinders are pistons (P) attached to connecting-rods (R), the lower ends of which are all fastened to a crank (X) on the shaft.

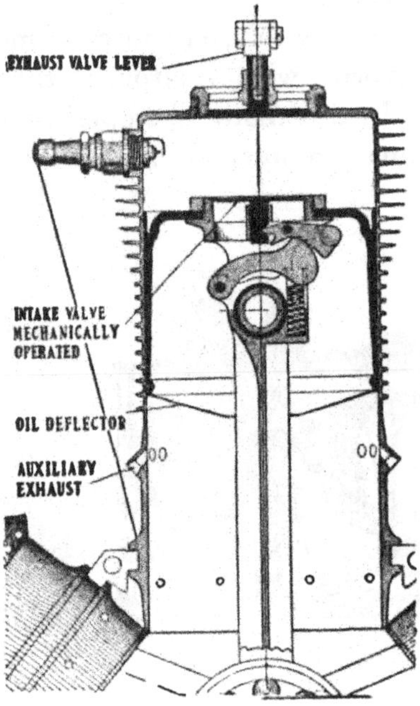

*Fig. 16*

By the explosion of a charge of gas in the cylinder (1) the force compels the cylinder, with the case attached, to travel *away* from the crank, as the crank itself is immovable.

As the cylinders and case revolve, the crank, being eccentric to the center, or shaft, pulls the pistons in cylinder 2 downward, thus drawing a fresh charge of gas into that cylinder. At the same time the piston in another cylinder (5) is being forced up to compress the charge which is then fired and forces the entire assembly around the shaft, and thus in rotation each cylinder fires its charge and helps force all the cylinders around the central shaft.

By reference to the diagram (Fig. 17) it will be seen that,

whereas the cylinders revolve around the shaft in the circle Y, the pistons and connecting-rods revolve around a center formed by the crank in the circle Z. In this way actual reciprocating, or up-and-down, motion is obviated, and the motor is free from jar and vibration.

The exhaust-valves are in the heads of the cylinders, as shown in Fig. 16, and are operated by push-rods, which are actuated by passing over stationary cams.

The intake-valve is situated in the top of the piston, and

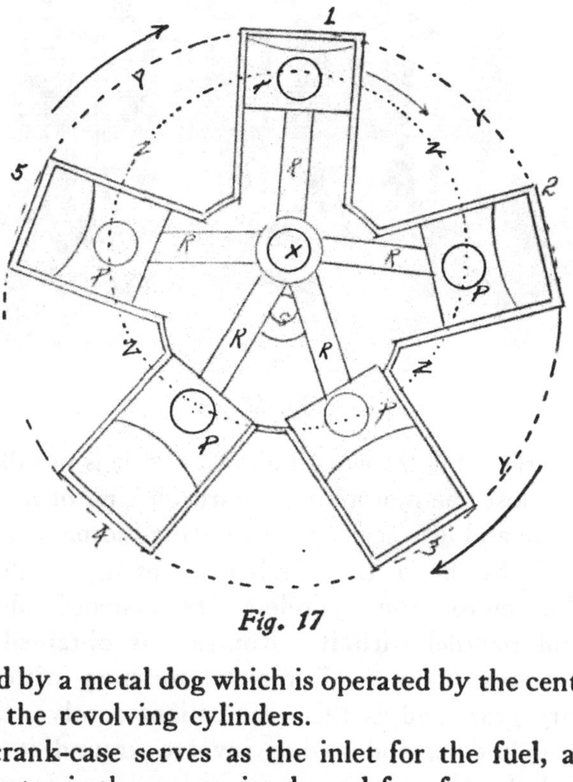

*Fig. 17*

is opened by a metal dog which is operated by the centrifugal force of the revolving cylinders.

The crank-case serves as the inlet for the fuel, and the whole motor is thus very simple and free from complicated parts.

These rotating-motors have been used extensively, and the majority of speed, altitude, and other records were made with rotating-engines. Unfortunately, they are very short-lived, excessively costly, and are very dirty in operation.

As they are lubricated by castor oil, and are whirling round and round at a speed of from 1,000 to 1,800 revolutions

*Fig. 18*

a minute, the oil is thrown all about, and it is usually necessary to inclose the motors in a guard, or rim, of metal.

A unique and distinct type of rotating-motor is the "Just-rite," made by the H. L. F. Trebert Company, of Rochester.

In this motor the cylinders are fastened about the shaft and parallel with it. Rotation is obtained by the pistons acting upon small pinions meshing with a large stationary gear, and as this gear cannot revolve, the small pinions, cylinders, and casing revolve around it and the central shaft.

In this motor there are no real valves, for each cylinder

receives and exhausts its charge by merely passing by openings or ports connecting with a chamber at the upper end of the motor. One half of this chamber is the exhaust, the other the intake, and as the cylinder which has just fired passes by the opening in the exhaust-chamber the burnt gas is expelled. It next passes the intake port and a fresh charge is drawn in.

Ignition is accomplished in an equally simple manner by means of a brush-contact wiping by a plate, and thus practically all wiring, a complicated timer, and other ignition apparatus are eliminated.

This motor has proved very successful, and, unlike most other aeroplane engines, it is so strong and powerful, and so free from ordinary troubles and breakdowns, that it has been used considerably in trucks and automobiles.

The only real advantage of the rotating over the stationary cylindered engine is the extremely light weight—frequently not over 2–1/2 pounds per horse-power—and the freedom from vibration.

For general use the stationary motor is far superior to any rotating-motor, and if constructed carefully and of proper materials, a stationary motor will prove far more reliable, longer-lived, and more powerful than a rotating-motor of the same cylinder capacity.

Both rotating and stationary aviation motors are made in air-cooled forms, and a great many of the stationary types are also made water-cooled.

Undoubtedly better results and longer life can be obtained by water-cooling, but the added weight of water, radiator, etc., is an objection.

All rotating-motors are air-cooled, and as the cylinders of these engines travel so rapidly through the air the air-cooling operates very well on this type.

Many excellent stationary-cylindered motors are also air-cooled, and both two and four cycle engines are made in air-cooled and water-cooled types.

The actual efficiency of either system is about the same; but, as a rule, an air-cooled engine is harder to lubricate, is more apt to be troublesome, and is shorter-lived than a water-cooled engine.

Both kinds are about equally used, and each builder and aviator has his own ideas on the subject.

Many of the most successful American motors are air-

*Fig. 19*

cooled, such as the Kemp (Fig. 18), while others which have proved excellent aviation engines are water-cooled. Among these are the Roberts (Fig. 19), a two-cycle type; the Curtiss, Wright, Sturtevant, Kirkham, Frontier, and

Mercury. The last is particularly interesting, as it practically is the only two-cycle V-shaped motor made (Fig. 20).

European makers strongly favor air-cooled motors, and the Anzani, Renault, R. E. P., and many others are of this type.

*Fig. 20*

Although so much time and money has been spent in trying to produce a reliable and efficient motor for aircraft, and such splendid workmanship and high-grade materials are used in aeroplane motor construction, yet the ideal engine is yet to be produced.

Probably no portion of the modern aeroplane is so prone to trouble and accidents as the motor, and, while wonderful progress in motor design and construction has been made, yet a great deal is yet to be accomplished before an aviator can feel absolute confidence in his engine, and may rest assured that his power will not fail at some critical moment.

## Chapter XIV

### MINIATURE AEROPLANES

IT is quite possible for any boy who is adept at using carpenter's tools and who has a fairly good idea of mechanics to build a real man-carrying aeroplane. The expense is, however, very great, and far beyond the reach of ordinary boys; and any boy who desires to fly, and whose parents do not object to his learning, will succeed far better by purchasing a standard aeroplane than by building one of his own.

A great deal of pleasure and much knowledge may be obtained by building exact duplicates of standard machines in model sizes, however, and if these working-models are carefully made they will fly splendidly.

It is very easy to build a copy of an aeroplane for show purposes from cardboard, flimsy materials, and odds and ends; but such models will seldom fly, and are of no particular value.

As the real biplane is easier to build than the monoplane, so the model biplane is easier than the model monoplane, and I will therefore first describe how to build a Wright biplane model.

Before commencing to build any of these little airships you should have the proper tools and materials on hand. If you have already built model fliers, the same tools you used will answer for building the model aeroplanes, and much of the same material may be used.

Fine drills, brad-awls, sandpaper, a small plane, cutting-pliers, and small, round-nosed pliers, soldering-tools, fine saws, and a tiny screw-driver are the most important tools; and strong silk thread, fine wire, glue, varnish and shellac, fabric, and spruce sticks and reeds are the most important materials.

The material for building a three-foot Wright model should not cost over six or seven dollars, even if bought ready cut, and if you get out some of the parts from rough stock yourself the cost may be considerably reduced.

In building a flying-model of a real aeroplane you must use a great deal more care than in constructing a model flier, for a very slight variation in size or proportion of parts will prevent the machine from flying at all.

It is always best to follow standard and tried designs at first, and after becoming familiar with details and principles you can try building models from your own plans or along original lines.

Many reliable firms, such as the Ideal Aeroplane and Supply Company, make a specialty of furnishing working-plans and materials for constructing scale-models of standard machines, and, as these models are guaranteed to fly if constructed according to plans, the boy builder cannot do better than to follow them.

The plans and directions given herewith are reproduced by permission of the Ideal Aeroplane and Supply Company, and if followed accurately any machine built in accordance with the plans will actually fly. It is useless to spend time, trouble, and money working at a model from unreliable plans, only to find upon completion that it cannot possibly rise from the ground.

## How to Build the Ideal Wright Biplane

The materials required for this three-foot model are as follows:

| | |
|---|---|
| 1 foot spruce 5/16″ square | 4 pieces 5/8″ aluminum tube 1/8″ inside diam. |
| 6 feet spruce 1/4″ square | 3 feet 1/4″ x 3/16″ spruce |
| 1 foot spruce 5/32″ square | 14 feet 1/4″ x 1/8″ spruce |
| 12 feet spruce 1/8″ square | 11 feet 3/16″ x 1/8″ spruce |
| 1 foot brass 3/32″ diam. (round) | 24 feet reed 1/8″ diam. |
| 1 1/12 yard silk fabric | 7 feet reed 3/32″ diam. |
| 2 10-inch Wright-type propellers | 20 feet tinned wire .32 gage |
| 64 feet rubber strands 3/32″ | 2 feet 1/16″ brass rod |
| Glue | 2 ball-bearing propeller-shafts |
| Model-making nails | 3 miniature wheels 1-1/2″ diam. |
| 1 11-inch special axle, 2 special aluminum tube braces | 7-inch aluminum rod 1/16″ diam. |

This will be ample for building the model, and will also allow quite a little extra material in case of breakage, and you must expect some breakage and waste in your first few attempts at constructing models from such fragile, thin material as is used in making these miniature machines. Before commencing actual operations you should study the plans (Figs. 1, 2, 3) very carefully and thoroughly. Fig. 1 is the ground-plan, or the view of the machine looking down upon it; Fig. 2 the front view; and Fig. 3 a section, or view from one side. The perspective view (Fig. 4) shows how the machine appears when completed.

Most of the important dimensions are indicated on the plan, but any part can be readily measured by means of the 12-inch-scale rule on the cut.

The various parts are all lettered, and by referring to the table at the top the name or significance of any part may be determined.

In building the model be very sure that each and every

# "IDEAL" WRIGHT AEROPLANE

## 3 FOOT FLYING MODEL

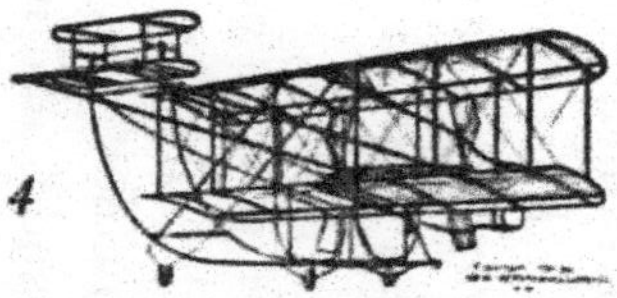

Fig. 4

| | KEY TO PARTS | | KEY TO MATERIAL |
|---|---|---|---|
| A | CHASSIS | a | Wood 5/16 Sq. |
| B | MAIN PLANES | b | " " 1/4 " |
| C | ELEVATION " | c | " " 5/32 " |
| D | RUDDER | d | " " 1/8 |
| E | PROPELLER | e | " " 3/32 Dia. |
| F | MOTOR | f | " " 1/4 x 5/16 |
| G | Ball Bearing Propeller Shaft | g | " " 1/4 x 1/8 |
| H | Rubber Tired Wheels | h | " " 3/16 x 1/8 |
| I | Special Axle | m | Reed 1/8 Dia. |
| J | Axle Brace | n | " " 3/32 " |
| K | Elevation Plane Axle | p | Tinned Wire |
| L | Aluminum Sleeves | r | Brass Rod |
| M | Rubber Skein | s | Silk |

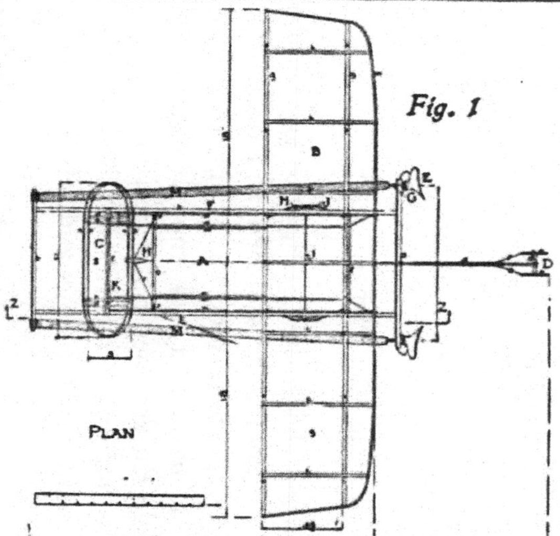



Fig. 1

PLAN

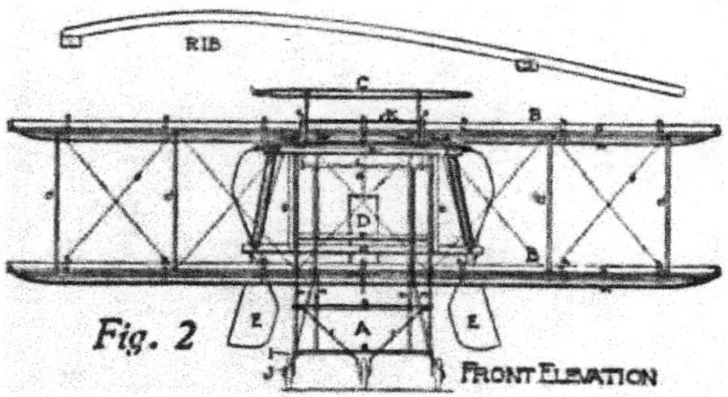



Fig. 2 FRONT ELEVATION

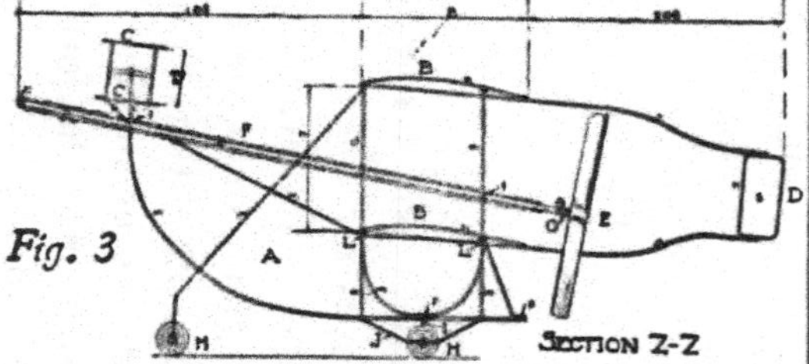



Fig. 3 SECTION Z-Z

joint is rigid and made with good glue and model-nails, and is wrapped with silk thread.

Never attempt to drive nails through small wood unless a hole—a trifle smaller than the nail—is first drilled (with a No. 60 drill), or otherwise the piece will split.

### Constructing the Planes

The 3/16″ x 1/8″ wood should be cut up into eight-inch lengths for the ribs, and should be steamed and bent into shape as follows: Copy the proper curve, as illustrated in the plans, on a piece of board, and tack enough small nails along the outline to hold the steamed rib in position. After steaming or soaking in hot water the rib may be bent and slipped into the form made by the nails and left until perfectly dry.

The pieces of 1/8″ reed for the wing outlines, as well as those for the chassis frame, should also be steamed and bent to shape.

The first thing to do after the ribs are steamed and bent is to construct the main planes, or wings. First bevel the front of each rib where it lies on the forward spar, and cut a notch in each rib where it crosses the rear spar, as shown in the diagram.

Place the ribs along the spars in their proper position and mark each spar at the points where the ribs cross, and mark where there are to be holes for stanchions, legs, and braces. Fasten the ribs in place by glue and a nail in each and lash with silk thread.

Place the frame with ribs in position on an outline (full size) of the plane, and trim off the ends of the ribs and spars to conform with it.

Next take the pieces of rattan, bent to the wing outline,

and place them around the spars and ends of ribs, fastening one end to the end of the front spar and carrying it around the rear ends of the ribs to the other end of the front spar. Fasten the reed edge to the spars and rib ends with glue and lay aside until thoroughly dry.

When the two planes have been thus completed in skeleton form, cover the upper surfaces with silk as follows: Coat the front edge of each plane with glue and place it on the silk and turn the latter up and *over* it. Next glue the reed rear edge, and stretch the silk over this edge and *under* it. Then coat the ends of the reed frame with glue and stretch the silk tightly across and under it in the same manner. When thoroughly dry, trim off the uneven edges close to the frame with a pair of scissors. If the silk, when completed, is coated with Ideal Coating Solution or some similar compound better results will be obtained.

After the upper and lower planes are finished and covered with silk the ends of the uprights, or stanchions, are trimmed down to fit in the 1/16″ holes which should have been previously drilled where marked. The four stanchion-holes in the central part of the lower plane should have 1/16″ aluminum sleeves placed flush with the tops of the spars, and into these sleeves the ends of the four stanchions (those of the rear being reed) are glued. After the glue is thoroughly hard cross-truss the spaces between the stanchions with wire, and the planes will be complete.

### Making the Chassis

The chassis, or frame, should first be made as a flat frame from two 27-inch pieces of reeds connected by the four cross-pieces (previously bored with 1/8″ holes); mark these W, X, Y, Z. The forward piece (W) is 7–1/2″ long, as well as the rear piece (Z). Glue the forward ends of the reed pieces into the outermost holes in W, and join them together at the points X, Y, and Z. When the glue is dry connect the frame to the main plane with four reed legs from holes in the lower plane-spars to holes in Y and Z. Connect X to the front spar of the upper plane with 12–1/4″ reed braces, and thus bend the chassis partly to shape.

Glue the forward ends of the 23-inch upper reeds into the innermost holes in W, and by drawing this back and fastening under the spars of the lower plane draw the chassis into the proper form and curve. Where this reed passes under the lower plane it should be wired to the legs, and at the rear of the back leg it should be bent down and finally wired and glued to the lower reed ends at *p*.

Through the aluminum sleeves in the lower plane-spars pass the ends of the semicircular braces (*m*), and wire the middle of each to the lower reed as shown. Before wiring these be sure and place the wheel-axle between.

Now place the wheels on the axle, screw the special axle-braces (*j*) to Y and Z and screw the nuts on the axle ends. Place the front wheel at the center of the 1/16″ brass rod, and bend the rod to the shape shown in the cut Bend the ends of the rod over X and the lower reed, and wire them firmly to X. Adjust this front axle so that the planes stand at about the angle shown in the section Z-Z, and the frame and planes will be complete and ready for the front elevator-plane.

### The Elevator

Make two H-shaped frames by joining two 3/32″ (round) uprights to a cross-piece 3/16″ x 1/8″, which must be drilled to rotate on a shaft. Bend the reed edge piece to the shape shown and join the edges with ribs, two in the upper plane and four in the lower. Cover the planes with silk, and leave spaces on the lower plane between the adjacent ribs where the shaft-support from the chassis passes through.

Attach the two planes to the H-shaped frames by gluing the uprights into holes drilled in the ribs, and when dry mount the whole on the shaft-supports by means of the aluminum shaft. Drill a fine hole in W, pass a wire through, and fasten the ends to the lower plane of the elevator. This will serve to maintain the elevator in any desired position.

Glue the 5/32″ square elevator-plane shaft-supports into the holes already bored in W, and the machine will be practically complete, and only requires a rudder, the rubber motor, and propellers to be ready for a flight.

### The Rudder

This is easily made by constructing two simple reed frames with the ends spliced together and made from the 3/32″ reeds. Each of the frames is then covered with silk and connected at top and bottom by cross-pieces of 1/8″ wood. The rudder-supports of 1/8″ wood are joined to the frame under the central ribs of both main planes, and the rear ends are nailed to the rudder cross-pieces. A small piece of wire passed through a fine hole in the upper support and attached to the front edges of the rudder frames will hold the rudder in any position it may be placed.

### Propellers

The propellers should be 10 inches in diameter and two inches wide, of regular Wright type, and they may be whittled from blanks or purchased ready-made. The latter method will insure better results, as the two propellers must be exactly alike, but right and left handed, so as to revolve in opposite directions. The propellers are to be mounted on the ball-bearing shafts and secured by nuts and washers, and by nails to the hubs.

The motor frame is next constructed by joining two long sticks by one intermediate and two end braces, as shown in the plan; but the forward piece should not be fastened in place until the frame is passed between the planes from the rear and is fastened in place at *ff*. Drill holes in the rear cross-piece, and in these place the shanks of the propeller-shafts and bend over the inner end of the shafts to form hooks. Drill two more holes in this piece—about three inches from the shaft—and in these place brass-rod stops to prevent the propellers from unwinding until the model is launched.

Divide the hank of rubber into two parts, and from each length form a skein and place one end over the forward cross-piece (*b*) and the other over the propeller-shaft hook.

The skein is easily made by winding the rubber about two nails placed 27 inches apart, so that seven loops are formed, and then knot the ends together.

When the rubber and propellers are in position turn each propeller about 250 revolutions, being careful that each is turned exactly the same number of times, and also being sure to turn them in opposite directions. The model is now ready to fly, and you will be very anxious to test it. After the time and trouble spent in making the dainty little Wright you should be most careful not to break it in the first attempt, and so you must select a very smooth, level, soft spot in which to try its flying powers.

Tilt the elevator upward slightly, set the tail true and straight, and release the stops from the propellers.

If you have followed directions carefully the little biplane will run a short distance on the ground and will then rise gracefully in flight and sail away for 100 feet or more, gradually slowing down and descending softly on the earth.

The first attempts should be made with a very slight elevation to the elevator in order to avoid mishaps, and the elevation may be gradually altered until the best results are obtained.

Instead of starting from the ground the model may be started from a board or other smooth surface tilted slightly upward in the direction of flight and placed about six feet from the ground. The machine flown in this way will fly farther and better, but is not as natural or as much like a real aeroplane.

After succeeding in constructing the Wright model you will be anxious to try other forms, and perhaps the best of these to work at next is the Nieuport monoplane.

### The Nieuport Monoplane Model

The materials used to construct this beautiful and successful model are as follows:

1/2 foot wood, 5/16″ square
6 feet wood, 1/4″ x 1/8″ square
15 feet wood, 5/32″ square
12 feet wood, 1/8″ x 1/16″ square
1-1/8″ wood, T section, 5/16″
2 feet wood, 3/32″ diam., round
2 feet reed, 3/16″ diam.
3 feet reed, 1/8″ diam.
12 feet reed, 3/32″ diam.
Nails; 1/4″ screws; eyelets; escutcheon-pins, and pins
1/16″ board 6″ wide
Ball-bearing propeller-shaft
Aluminum propeller-hanger
12″ wooden propeller
10 feet para rubber 1/8″ square
2 sheets bamboo-paper
2 oz. bamboo-varnish
Set of 12 Nieuport sockets
1 Nieuport running-gear with axle
2 rubber-tired wheels 2″ diam.
1/16″ aluminum rod
1/16″ steel rod
Some thin sheet aluminum
Spool of No. 30 wire

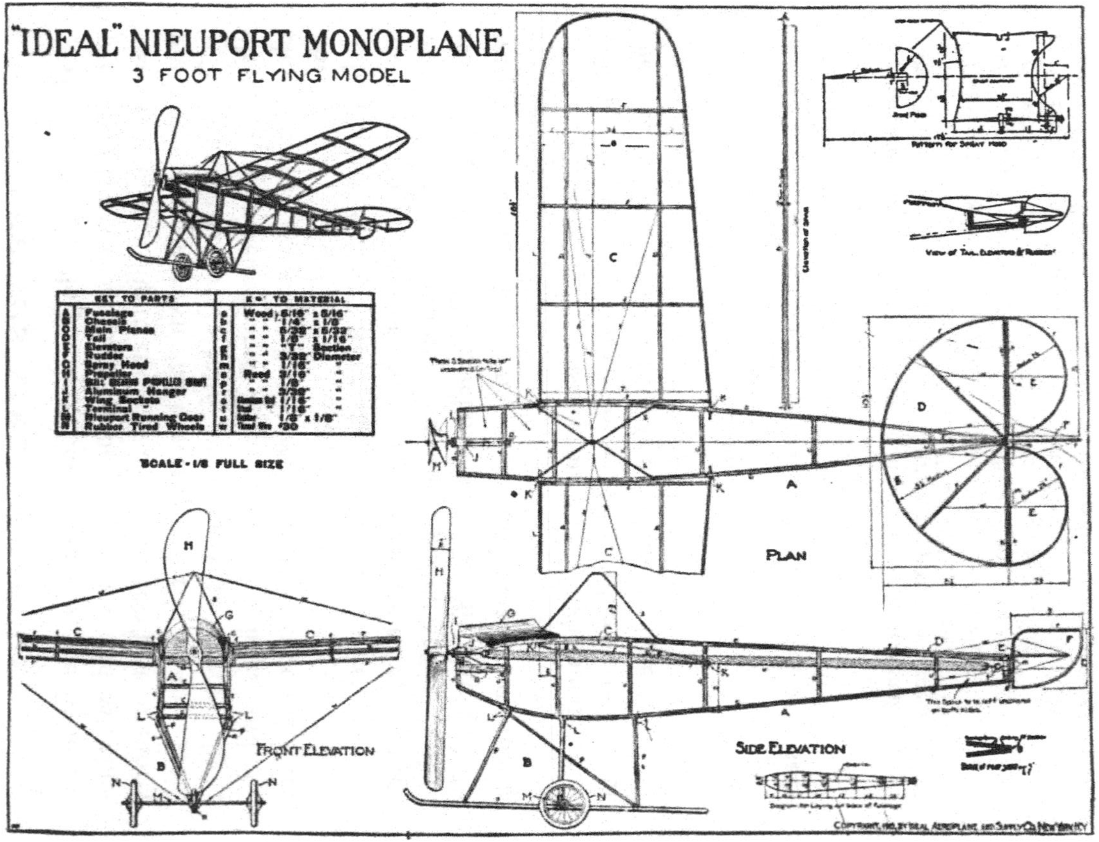
"IDEAL" NIEUPORT MONOPLANE
3 FOOT FLYING MODEL
KEY TO PARTS
Fuselage
Chassis
Main Planes
Tail
Elevators
Rudder
Spray Hood
Propeller
Aluminum Hanger
Wing Sockets
Terminal
Nieuport Running Gear
Rubber Tired Wheels
SCALE - 1/8 FULL SIZE
PLAN
FRONT ELEVATION
SIDE ELEVATION

The above should not cost over six dollars, and all the materials may be bought ready cut to size for five dollars; but if you purchase the materials in various lengths and cut them yourself you are less likely to run short through breakage, and can use what you have left from one model in building others.

The plans of the machine, as illustrated, show the details, dimensions, and methods of construction so clearly that little explanation is required.

The same care should be used in building this model as in the Wright; the joints should all be strong and rigid, no nails should be driven without first boring holes, and the ribs, four fuselage-bars, and the reed-outline pieces should all be steamed and bent exactly as described for the Wright biplane model.

### Constructing the Fuselage, or Body

First make the sides and join the top and bottom bars, after forming, to the vertical struts, thus making a frame as shown in side elevation. Join these two frames with horizontal struts, as shown in plan, and be sure that the distances between frames, struts, etc., are *exactly* as illustrated.

Through the T-section piece drill two holes to take the rear hook (side elevation); bend the hook into place, and join the T-section to the rear of the body, as shown in the detailed view of the rear joint (view of tail and rudder).

Join the other 5/16" piece to the 5/16" strut which belongs at the front of the body, and over both screw the aluminum propeller-hanger.

Place the ball-bearing propeller-shaft in this, and bend the inner end of shaft in the shape of a hook, as shown in side elevation. Attach this bed and hanger to the body frame, and also fasten it to the extra 5/32″ horizontal strut (side elevation), being very careful that the shaft is in a straight line with the rear hook. The ends of the 5/16″ horizontal strut must be cut away until the front comes flush with the front of the fuselage; then wire it very firmly to the fuselage.

The wing-sockets should next be screwed in position on the frame of the body as shown in side elevation. Next cover the fuselage—except where indicated—with bamboo-paper, and give it a coat of bamboo-varnish.

The tripod should now be constructed of 1/16″ aluminum rod, wiring the tops together and flattening the ends, and then wiring these to the upper fuselage-bars (side elevation).

The propeller may now be placed on the shaft and fastened with the nut and washer and pinned through the flange by escutcheon-pins.

Knot the ends of the rubber and divide the loop thus made into six loops; hook one end of the skein so formed to the hook on the propeller-shaft, and the opposite end over a small S-hook made from the steel rod.

Hold the fuselage front end up and allow the rubber to drop down to the rear end, where it may be readily hooked over the rear hook through the opening left in the body covering.

### The Chassis, or Running-Gear

Make the curve on one end of the skid, as shown in side elevation, and place upon it the ready-made Nieuport running-gear, and glue it in position. Drill holes near one end of each of the five 1/8″ reed braces and wire them to the skids as shown, and insert the 3/16″ reed braces in the sockets of the running-gear.

To the upper ends of the braces glue the seven terminal sockets, and screw all to the fuselage before the glue is dry, and then true it up.

After the glue is dry insert the axle in the running-gear and mount the wheels.

### The Spray-Hood

On the sheet aluminum draw the pattern of the hood, as detailed in plan of spray hood, and cut it out carefully with shears. Draw the front piece on thin wood and cut it out, being very careful not to split it. Wire the aluminum hood to the front piece through holes drilled through each, and fasten it in place over the propeller-shaft as shown.

### Making the Wings

Trim down the 1/4″ x 1/8″ pieces so that they are 1/4″ high at the fuselage and 3/32″ at the tips, and to insure even tapering mark them off first with a straight-edge.

Wire the ribs, already bent, one above and one below the spars in the positions shown in plan. Then cut off the rear ends of the ribs to the proper size and form. Next wire the reed frame of the wing—which is already bent—to the ends of ribs and spars through holes drilled for the purpose.

The ends of this outer piece of reed should project 5/8″ at the body end, so as to fit into the wing-bars on the fuselage. In making the wings do not forget that the *right* and *left* wings are *different*, and they *must* be made *right* and *left*.

Make small hooks on the ends of four pins and pass the other ends through holes drilled in the spars, and then bend hooks on these ends. To the hooks thus made the wires which run to the tripod and chassis are to be attached.

Now glue all the joints, true all up, and hang the wings up to dry. When thoroughly dry the wings should be covered with bamboo-paper, the under sides being covered first, and be sure that the paper adheres firmly to each rib as well as to the reed-frame piece. When covered and dry give the paper a coat of bamboo-varnish.

Make four S-hooks from pins, and hook two on the top of the tripod (front elevation), and the other two on the skid just back of the running-gear.

Slip the projecting ends of the rattan wing frames into the sockets on the fuselage and tie wires to the hooks in the wing-spars, and carry them to the S-hooks on the tripod and running-gear, inserting little eyelet tighteners or tiny turnbuckles, as shown in front elevation.

Tighten the wires so that the main planes, or wings, tilt slightly upward as shown in front elevation. The wings are easily disassembled by merely unhooking the S-hooks from the skid and tripod.

### The Tail

Wire the semicircular outline of reed to the 3/32″ round wood, through holes drilled in the reed, and fasten the three radiating-ribs in the same way. Glue them together and dry between two flat surfaces to insure perfect truing up. Cover the framework thus made with bamboo-paper and coat with bamboo-varnish.

For the rudder, wire the ends of the reed, bent in shape as shown in side elevation, glue them together, dry between flat surfaces, and cover the same as the tail.

Construct the elevators in the same manner as the tail, and when these and the rudder are dry the 1/16″ tiller-bars are bound and glued in position as shown in plan, side elevation, and detailed plans. Nail the tail to the top of the fuselage by two nails through the central radiating-rib, *being sure to drill holes first.*

Next hinge the rudder to the rear T-piece through holes drilled in the latter, and tie it in place with wire, using the tiller-bars as shown. This will maintain the rudder and elevators in any position desired.

The three-foot model will now be complete and ready to fly, and if you have done your work well you will have the prettiest, most graceful, and strongest little aeroplane you ever saw.

It will prove a splendid flier, and may be flown from the ground or launched from the hand.

In flying release the propeller first, and then let go of the machine; and if you wish the model to rise rapidly raise the rear end from the ground, so that the skid will not drag, and at the same time give the machine a slight forward thrust.

In launching from the hand grasp the underside of the body just behind the chassis with one hand, and hold the propeller with the other hand.

Let go of the propeller and gently thrust the model forward, and keep it on as nearly a level keel as possible.

At first you should only try flights on calm days, or in a gentle wind, and in perfectly open and smooth spots.

If you wish a larger model of the Nieuport, or, in fact, of any of the machines described, you can readily construct them by enlarging the various parts and drawing them out to scale, following the plans given for details and proportions. In this way you can build models up to six feet spread if you choose.

Still another model reproduction of a standard type of

aeroplane which you can easily build is the Bleriot, and this is such a well-known and famous machine that you should certainly add one to your fleet of model flying-machines.

### The Bleriot Model

If you have followed the instructions and built the Nieuport and Wright machines, you will find the plans of this model so simple and easily understood that you will have no difficulty in following them, for the principal dimensions are given, and the reduced one-foot rule will serve for measuring the other parts.

The materials and tools required are practically the same as for the other models; the dimensions of wood, etc., as well as special fittings, being as follows:

4 feet wood 5/16" x 5/16"
2 feet wood 1/4" x 3/16"
7 feet wood 5/32" x 5/32"
20 feet wood 1/8" x 1/8"
12 feet reed 1/8" diam.
7 feet reed 3/32" diam.
1 foot aluminum rod 1/16"
1 foot brass rod
Silk for covering planes and rudder
Two rubber-tired wheels 2 inches diam.
Ball-bearing propeller-shaft
Wooden propeller 12 inches diam.
Aluminum propeller-hanger
40 feet para rubber 1/16" square
Model-making nails
Spool No. 32 wire
Small piece of sheet aluminum
18 smallest screw-eyes
4 small screws
4 tiny tube turnbuckles

These will cost about $3.50 to $4, or the parts may be bought ready-cut to lengths and size complete for $4. I advise all boys who wish to build several models, or who are likely to carry out their own ideas, to buy all wood, reed, etc., in quantities, and thus have a stock on hand which will save delays in ordering more. For boys who wish to build but one model the ready-cut wood and reed, with attendant fittings selected and packed especially for making each particular model, is the best.

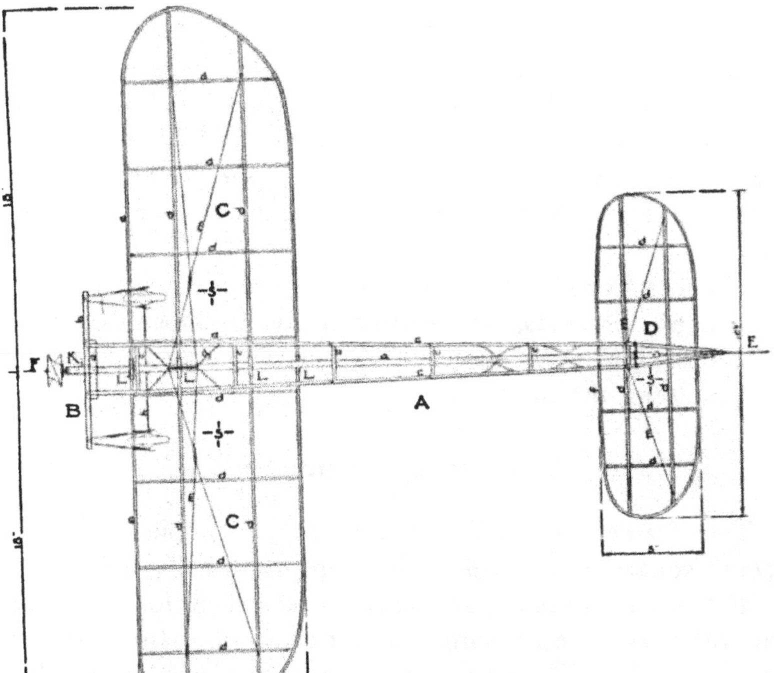


"IDEAL" BLERIOT MONOPLANE. THREE-FOOT FLYING MODEL

In the Bleriot model the joints must be strong and rigid, perfectly true, glued and nailed—save where reed and wood lie parallel, in which case glue and No. 32 wire should be used. It is always a good plan to lash all joints neatly with silk thread and finish by coating with shellac or Ambroid varnish.

The first step is to shape the wooden ribs and curved reeds for wings, tail, skids, etc, as directed for the models already described, and while these are drying you may busy yourself building the fuselage.

### The Fuselage Construction

Two frames are made from four 32-inch pieces of the 5/32″ square wood, two of the long pieces being joined by 5/32″ square uprights, as shown in elevation, for one side; the other two pieces being joined exactly the same way for the other side. These two side-frames are then connected by six horizontal pieces of the same wood on top and six on the bottom, as shown in plan.

In order to avoid two nails coming close together in the main-frame beams the horizontal cross-pieces should be placed slightly in the rear of the upright pieces, as illustrated.

The T-section tail-piece (*r*) is next placed in the narrow end of the frame, trimmed flush at top and bottom, and the body is complete.

The aluminum rod should now be bent to the form shown at *g*, elevation; two pieces exactly alike being made, the central part lashed together and the ends twisted around the fuselage-bars and lashed in place.

The tripod beneath the body (*g*, elevation) should be bent and lashed in the same way, and in making and attaching these you should be very careful to see that the highest point of the tripods is exactly in the center of the fuselage from side to side.

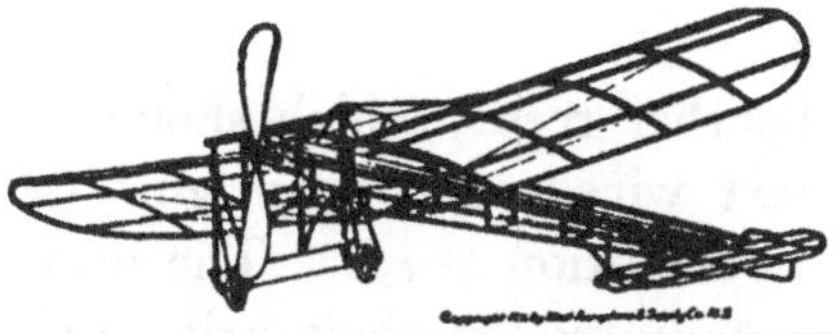

| | KEY TO PARTS | | KEY TO MATERIAL |
|---|---|---|---|
| A | Fuselage | a | Wood [illegible] |
| B | Chassis | b | " [illegible] |
| C | Main Planes | c | " [illegible] |
| D | Elevation " | d | " [illegible] |
| E | Rudder | e | Reed [illegible] dia. |
| F | Propeller | f | " [illegible] " |
| G | Shock Absorber | g | Aluminum Rod |
| H | Skid | h | Brass " |
| I | Motor | m | Tinned Wire |
| J | Ball Bearing Propeller Shaft | n | Rubber Strand |
| K | Aluminum " Hanger | o | " Tie |
| L | " Sleeves | -ɸ- | Silk |
| M | Rubber Tired Wheels | r | Special Tail Piece |

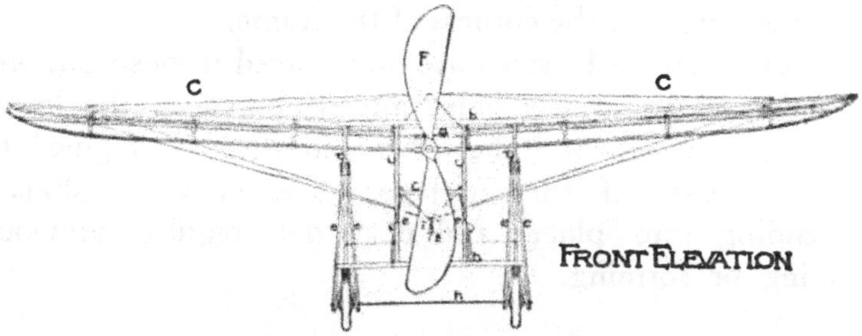


"IDEAL" BLERIOT MONOPLANE. THREE-FOOT FLYING-MODEL. DESIGN "B"

### Making the Chassis

The frame of the chassis is made from two 1/4" x 3/16" pieces drilled near the ends with 1/8" holes for the uprights, and with two other holes in line with the fuselage sides to receive the braces (*e*, elevation). Join the two pieces with 1/8″ x 1/8″ pieces, the ends of which should be trimmed down to fit the holes. Two other vertical pieces and a short horizontal piece of 5/32″ square are made into an inner frame, which is joined centrally to the main frame as shown.

The shock-absorber reeds, which should have been bent to form, are next wired to the uprights (*d*), one at each side, as shown in elevation at *e*, *e*. The rubber-tired wheels should then be mounted on their axles and placed in the forks formed by the reed absorbers, the axle being secured to the inner prongs of the forks by wire, and to the outer prongs by bending the axle-ends around the reeds and clinching them with pliers. A great deal of care should be taken to have the axles true and the wheels parallel, so that the wheels will run smoothly.

The ends of the long fuselage-beams are next joined to the inner frame of the chassis, care being taken to have the ends fit snugly into the corners of the frame.

The diagonal reed braces are now placed in position, one end being glued in the holes already bored in the lower bar of the chassis, and the other end wired and glued to the lower bars of the fuselage. The curve is shaped by bending into place, and does not require previous steaming or forming.

### The Skids

These are of reed, 3/32″ diameter, formed as shown in elevation, H, and tied with wire where they cross below, and fastened with wire and glue to the lower bars of the fuselage. If a small 1–1/2″ rubber-tired wheel is attached to the skid of this model the machine will rise quicker from the ground and will be greatly improved.

## The Motor

Cut a 1/16″ groove in the small 5/16″ square cross-piece, and into this fit the motor-stick (*a*). Fit on the aluminum propeller-hanger tight against the stick and fasten with screws, the upper part being bent over the cross-piece.

Insert the ball-bearing shaft through holes, and bend the inner end into a hook; string the rubbers between this hook and another at the rear end, which is fastened through a hole, and wind the rubber back and forth, keeping it rather tight, until a skein is formed. The ends of the rubber should then be tied to the hooks. The motor is now complete, and should be fastened in the fuselage by nails and glue in front, and by a piece of rubber tied to the upper fuselage-bars in the rear, as shown in elevation and plan. By this method of attachment all twisting of the fuselage by the motor is avoided.

The propeller should next be fastened to the shaft by a nut, with nails through the hub, and the body laid aside while you make the wings and tail.

## The Main Planes

The bent ribs are joined to two beams, or spars, of 1/8″ x 1/8″ wood, and the ends trimmed off to conform with the outline shown, and around the ends of the ribs and bars the reed-outline piece is bent and fastened.

In making the wings bear in mind that they are ***right and left***, as noted in the directions for building the Nieuport model.

The planes should be covered with the silk on the upper sides by giving the reed edges a coat of glue, laying it on the silk, drawing the silk tightly over the reed, turning the edges over, and finally trimming off with scissors.

The two separate wings thus made are joined together through the fuselage by tight-fitting aluminum sleeves, or ferrules, and are wired in position by stays of wire passed from screw-eyes in the spars to the aluminum mast on the top of the body and to the tripod below. By using the tiny turnbuckles on these wires far more accurate adjustment can be obtained than by any other method; but a small piece of tube slipped over the wires as shown in plan will serve the same purpose.

By noting the spot where the reed frames and spars pass through the fuselage, and the distances from the cross-bars at these points, you can readily determine just the angle to give the wings.

### The Rudder and the Elevator

The elevator is made in the same way as the main planes, except that it is in one piece and the ribs are not curved. While the glue is still soft the frame should be placed between weighted boards to dry, and thus insure it being perfectly flat and true.

It is covered with silk, as in the main planes, and is attached by the front spar to the fuselage by nails, and is wired in position by stays to screw-eyes set in the sides of the upper fuselage-beams. The angle can be varied to obtain various kinds of flights.

The construction of the rudder is similar to that of the planes, except that it is covered with silk on both sides, and the frame must be flush with the cross-bars on both sides by making half-and-half lap joints. It is hinged to the tail-piece of the fuselage.

The model is now complete, and will prove to be an exact copy of a Bleriot monoplane, with all the grace, birdlike form and easy flight of the large machines. It will rise from the ground and fly 100 feet or more, or may be started from an upward-inclined surface as directed for the Nieuport monoplane.

Having constructed these three models, the boy model-maker will be perfectly familiar with methods and details, and with little trouble can draw plans of any standard machines and from them build models of various kinds.

Most of the aeronautical magazines, such as *Aeronautics*, *Aircraft*, etc., publish scale working-drawings of various aeroplanes in each issue, and by reducing these to model sizes it is very easy to build the models.

Models of the latest types of Curtiss, Moissant, Deperdussin, Morane, Farman, Caudran, Baldwin, Martin, Burgess, Benoist, Thomas, and scores of other makes may be built, and if you can form a club of boys and build the models together a great deal of fun and a wonderfully valuable training in aeroplane construction and design will be obtained.

# Part V

# HYDROAEROPLANES AND FLYING-BOATS

# Chapter XV

## THE HYDROAEROPLANE

ODDLY enough, the most recent type of aircraft, the hydroaeroplane, is really the oldest form of aeroplane. The first experiments of Voisin, Bleriot, etc., made as early as 1905 and 1906, were conducted on water. This was done as a matter of safety, and not with the idea of developing the hydroaeroplane, for just as soon as any real flights were obtained wheels were attached to the machines and the craft became aeroplanes.

It was not until several years after the Wrights and others flew successfully that inventors turned their attention to the possibilities of the hydroaeroplane, and in a short time

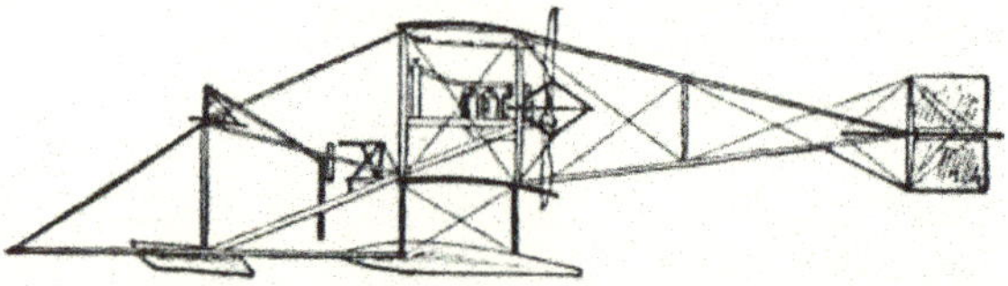

*Fig. 1*

practical machines of this sort were produced. The first American hydroaeroplane was that of Mr. Curtiss (Fig. 1), and to-day the Curtiss hydroaeroplanes and flying-boats are among the most efficient in the world.

Rapid as was the development of the aeroplane, the hydroaeroplane progressed even faster; and within two years these machines have become great favorites, with wonderful records to their credit.

A Benoist hydroaeroplane flew from Omaha to New Orleans, a distance of 1,835 miles, while a Curtiss army type has made a non-stop flight of six hours over water—the world's record for 1912.

The word "hydroaeroplane" is often confused with "hydroplane," but in reality the two are very different. The real *hydroplane* is a form of boat which is provided with jogs, or "steps," on the bottom, which cause it to rise upward when in motion and skim over the surface of the water instead of going through it. Hydroplanes are the fastest of boats, and high-powered racing-hydroplanes have traveled at a speed of over fifty miles an hour.

The hydroaeroplane is a combination of the hydroplane and the aeroplane; in other words, it is an aeroplane attached to a boatlike hull or to floats which skim over the surface of the water like the true hydroplane. Formerly all such machines were termed "Hydroaeroplanes," but the latest types, which have a single boatlike hull, are now more commonly known as "flying-boats" or "flying-yachts," and are so different from the ordinary hydroaeroplane as to be worthy of a class by themselves.

The advantages of the hydroaeroplane over the true aeroplane are many. In the first place, they are much safer—only one or two fatal accidents have ever occurred in these machines—for even if the machine falls there is little likelihood of any serious injury resulting to the machine or pilot.

Moreover, it is not necessary to fly high in a hydroaeroplane, for the steadiest and best air currents are usually close to the surface of the water, whereas over land the aviator must rise to a considerable height in order to avoid dangerous air currents.

In flying over water the aviator may descend and stop

his machine at any time and place, for the water presents a fairly flat surface at all times when calm, whereas over land the pilot must descend on a selected spot in order to avoid accidents, and frequently such level or open places are not at hand. In addition it is often very hard for an aviator to tell just what sort of country is beneath, for from a height brush, uneven land, and level turf look much alike.

One well-known airman, while flying over New York City, was obliged to descend, and selected a broad green expanse which he thought was level and open fields. After a long and successful glide his machine turned a complete somersault as it plowed into the long grass and reeds of the Hackensack meadows.

Only the soft marsh and the deep grass prevented the aviator from being killed, and, as it was, his machine was badly injured.

Many hydroaeroplanes have no means of landing on the earth; but the later models combine the floats for water with wheels for ground landing, and when thus equipped the aviator may fly over land or water and descend where he chooses, and may rise from either earth or water with equal ease.

### Pontoons, or Floats

The earlier aeroplanes that were equipped for water-flying had two floats, or pontoons, attached to the landing-chassis, and several standard machines still use these double hulls.

Such was the hydroaeroplane of Frank Coffyn, which was flown about New York harbor in the winter of 1912, and which attracted world-wide attention by its success and efficiency.

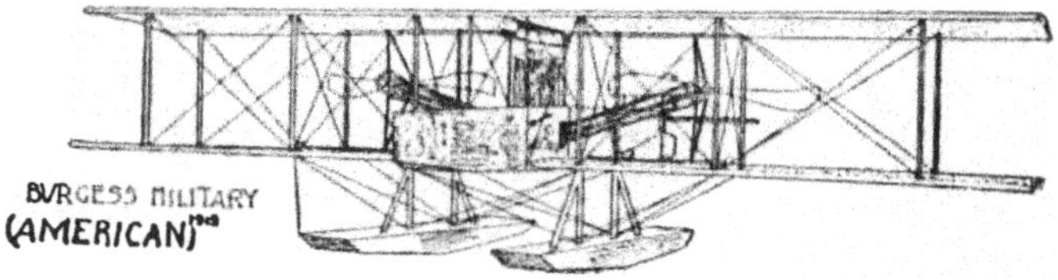


*Fig. 2*

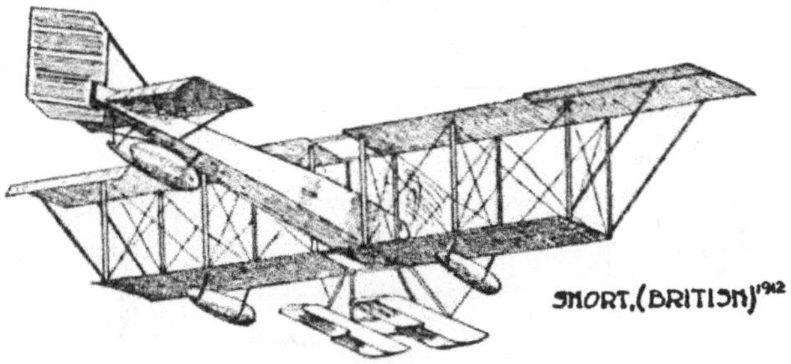


*Fig. 3*

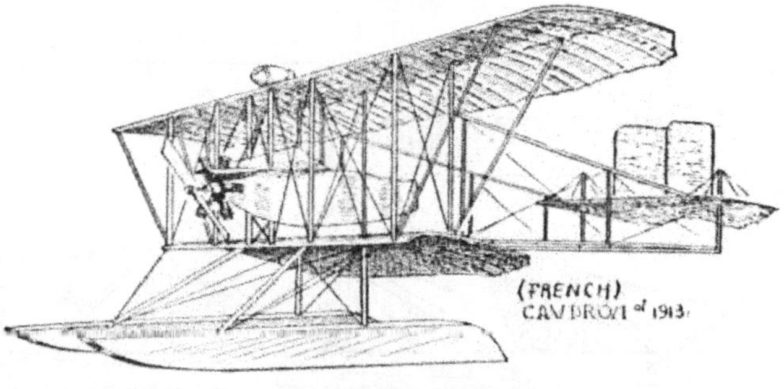


*Fig. 4*

The Burgess, the Caudron, the Farman, and many other machines also use two floats. Some of these are illustrated in Figs. 2, 3, and 4, while the details of the floats of the Curtiss, Burgess, and Coffyn machines are shown in Fig. 5. Practically all these pontoons are provided with steps on the bottom, like those on the hydroplane proper, and even when traveling on the surface of the water they attain a

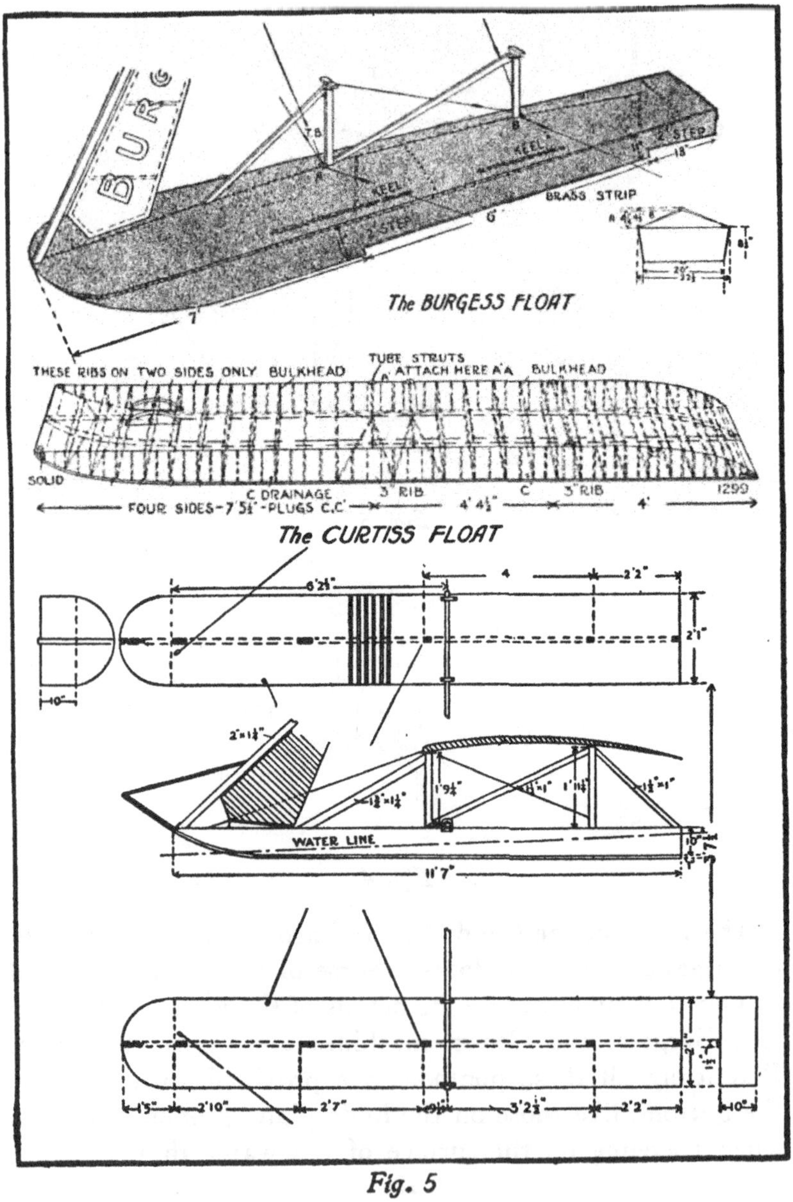


*Fig. 5*

**DETAILS OF THE FLOATS OF THE CURTISS, BURGESS, AND COFFYN MACHINES**

speed of 35 to 40 miles an hour, thus combining all the advantages and speed of motor-boating and flying.

Several builders soon discarded the double-float system and fitted their machines with a single hull, or float. Among these are the Curtiss, Benoist, etc. A view of the Curtiss

*Fig. 6*

machine on the water is shown in Fig. 6, and a larger view of its hull is illustrated in Fig. 7.

Aside from the float, or hull, this machine is very similar to the standard Curtiss aeroplane, and, in fact, the majority of hydroaeroplanes can be transformed to true aeroplanes by removing the hull and substituting wheels or running-gear.

The majority of hydroaeroplanes are of the biplane type; but during the past year many monoplanes have been equipped with hulls, and have proved very fast and efficient hydroaeroplanes, but as a rule they are more difficult to handle and are less reliable than the biplanes.

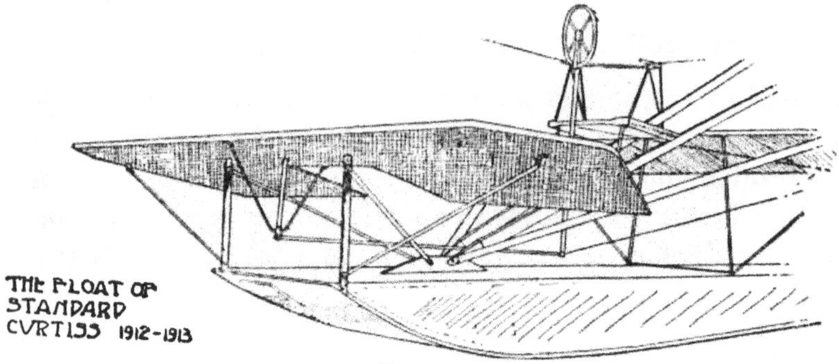


*Fig. 7*

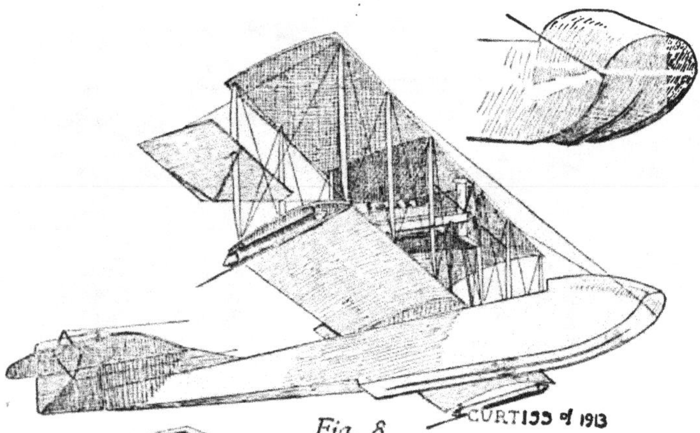


*Fig. 8*

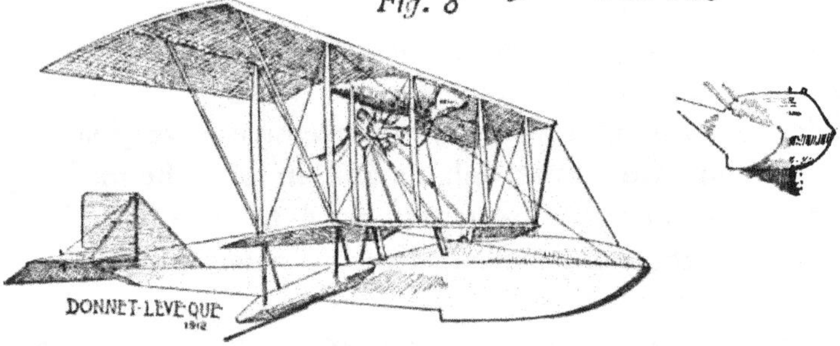


*Fig. 9*

## The Flying-Boat

With the perfection of the hydroaeroplane and the appreciation of its manifold advantages, builders at once turned their attention to this type of machine; and as the Navy Department of the United States drew up specifications for machines which would answer naval requirements, each builder strove to produce a machine which would be acceptable to the government.

This resulted in the creation of the most perfect and highly developed aircraft in existence—the "flying-boat," or "flying-yacht."

Strange as it may seem, the flying-boat appeared in France and America at almost the same time, and in forms and details so much alike that an ordinary person could not distinguish the differences.

The Curtiss flying-boat in this country, and the Donnet-Leveque in France, both came into prominence simultaneously, and both proved wonderful revelations in the possibilities of aircraft.

How similar these two flying-boats really are may be judged by comparing the illustrations (Figs. 8 and 9). It will be seen that the shape of the hull, the small cylindrical floats on the ends of the lower planes, the arrangement of the tail and the rudders, and the general appearance are practically identical. The Curtiss, however, has the standard type of Curtiss ailerons *between* the wings, whereas the Donnet-Leveque has hinged flaps, or ailerons, on the rear edge of the upper plane.

The Curtiss machine, shown in Fig. 8, is equipped only for water-flying, as is also the French machine, shown in Fig. 9; but both flying-boats are also built equipped with

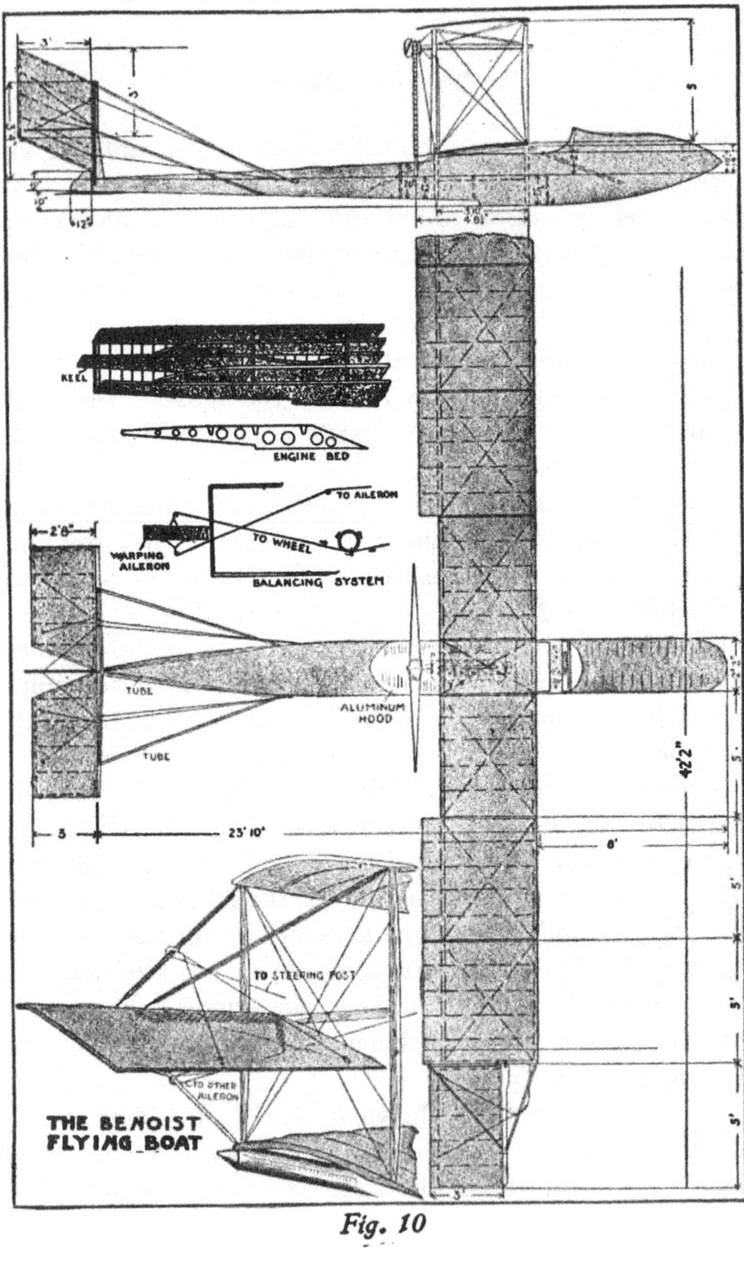
KEEL
ENGINE BED
TO AILERON
WARPING AILERON
TO WHEEL
BALANCING SYSTEM
2'8"
TUBE
ALUMINUM HOOD
TUBE
42'2"
23' 10"
8'
5'
5'
5'
5'
TO STEERING POST
TO OTHER AILERON
THE BENOIST FLYING BOAT

*Fig. 10*

wheels which can be folded up beside the hull, and can be dropped at will when the operator wishes to descend on land.

The success of these two types was so great that now a number of builders construct flying-boats of remarkable efficiency and very graceful appearance. Among these are the Benoist, which recently came into prominence by its splendid flight from Omaha to New Orleans over the Mississippi River, and plans of which are shown in Fig. 10.

Another interesting craft of this class is the *Miss Columbia*, illustrated in Fig. 11.

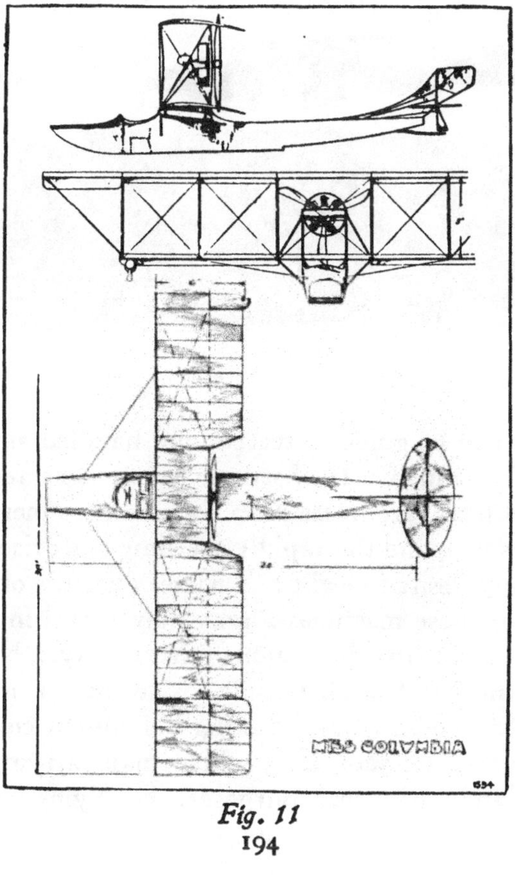


*Fig. 11*

The best known of all flying-boats in America is still the Curtiss, which was the original "flying-boat" in this country.

The Curtiss flying-boat is as safe and steady as a motor-boat, and even when traveling on the water at forty miles

*Fig. 12*

an hour it can be guided, steered, and handled as readily as an ordinary launch. If obstructions, such as reefs, boats, or other objects, are in its path the operator merely has to use his elevator, and the rapidly speeding craft rises from the water to any desired height and passes over the obstruction.

In one of these machines a man may travel in a straight line from place to place along rivers, bays, harbors, or lakes, for necks of land, islands, or towns do not prevent the flying-boatman from following his chosen course. For racing, sport, tenders to yachts, mail-carrying, and a thousand and one other purposes, the flying-boat is the most pleasurable, efficient, and fastest method of transportation ever devised.

When not in use it may be run up on a beach or float, or it may be tied to a stake or anchored like a yacht, or the wings may be quickly taken off and the hull and frame hoisted to the davits of a yacht.

The value of the flying-boat in naval warfare was quickly recognized by naval authorities, and the various navies of the world are now equipped with many craft of this type.

For scouting, message-carrying, bomb-dropping, and other services, the flying-boat is far superior to any other aircraft, and doubtless the naval war of the future will bring into action hundreds of these speedy and amphibious machines.

In flight the flying-boat looks more like some sort of sea-bird than anything else, for its tapered body, broad wings, and graceful flight are very different from the common aeroplane, and a good idea of its appearance may be gained by the photograph reproduced in Fig. 12.

Commencing as an aeroplane designed to rise from and descend on water, the hydroaeroplane has rapidly developed into a motor-boat equipped with wings, and at no distant date flying-boats may cross the broad Atlantic at a speed which will make the fastest ocean liners seem slow in comparison.

## Chapter XVI

### HOW TO BUILD A MINIATURE CURTISS HYDROAEROPLANE

IF the boy interested in aeroplanes has built a model of the Wright or any other real machine, he will have little difficulty in constructing a model hydroaeroplane; and if he wishes real fun and sport with his models he should certainly add one of these craft to his collection.

Fascinating as it is to see a model machine rise and fly from the ground, it is far more interesting and attractive to see a little model "skitter" across the water and finally arise from the surface and sail away in actual flight, and you and your friends will never tire of watching it.

It is best to commence with the hydroaeroplane proper; but after you have built this model you can readily change proportions and details and build a true flying-boat.

The machine described is a convertible hydroaeroplane, so when built you will have both a true Curtiss aeroplane and a standard Curtiss hydroaeroplane.

Owing to the floats, the numerous additional parts, and other details, the drawings appear far more complicated than in the other models described; but if you study them and read the directions carefully as the work progresses you will find no difficulty in making the model.

The first thing is to copy all the plans shown, A to H, enlarged to full size on a smooth board, or on several boards, being sure to mark every joint. Then cut the

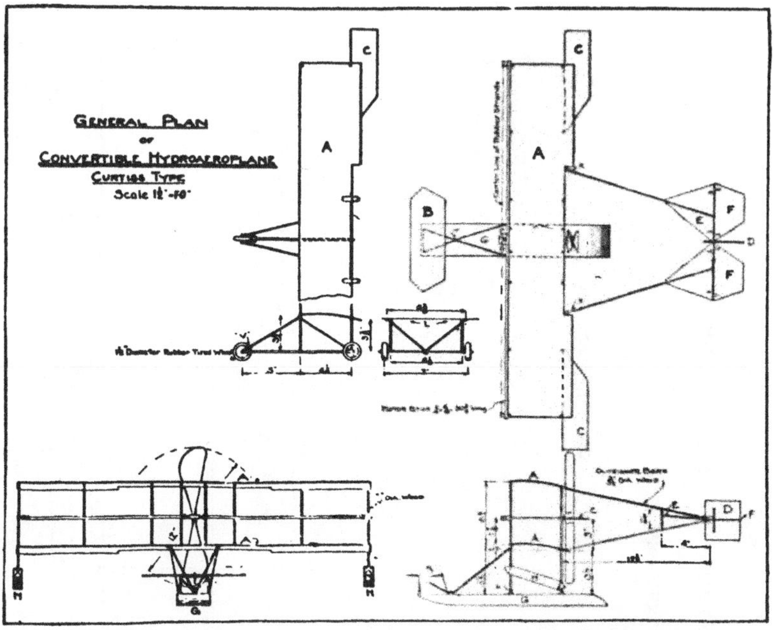




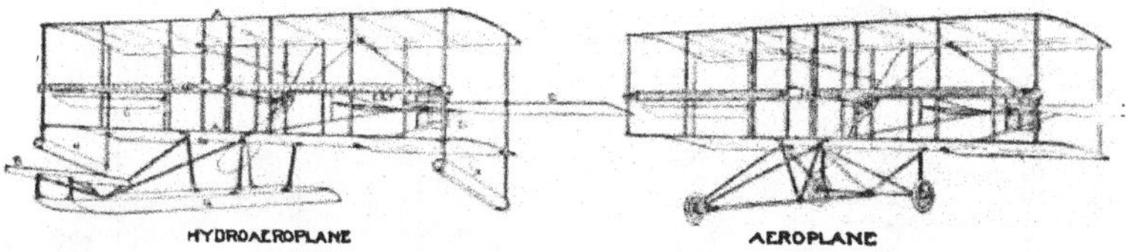


GENERAL PLAN

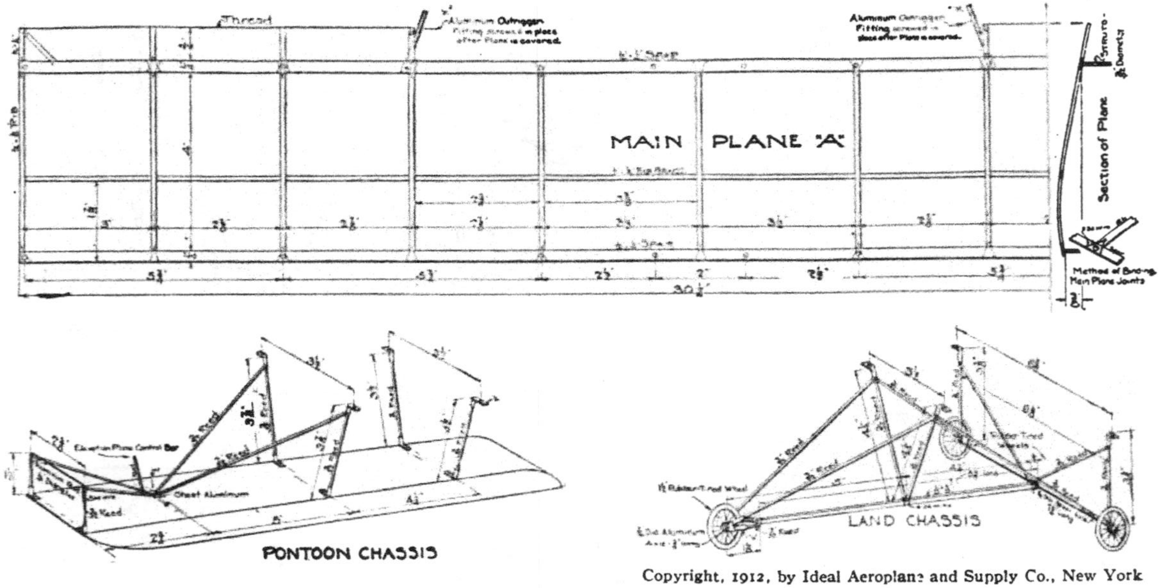


PLAN AND GENERAL VIEWS

"IDEAL" CURTISS HYDROAEROPLANE. CONVERTIBLE

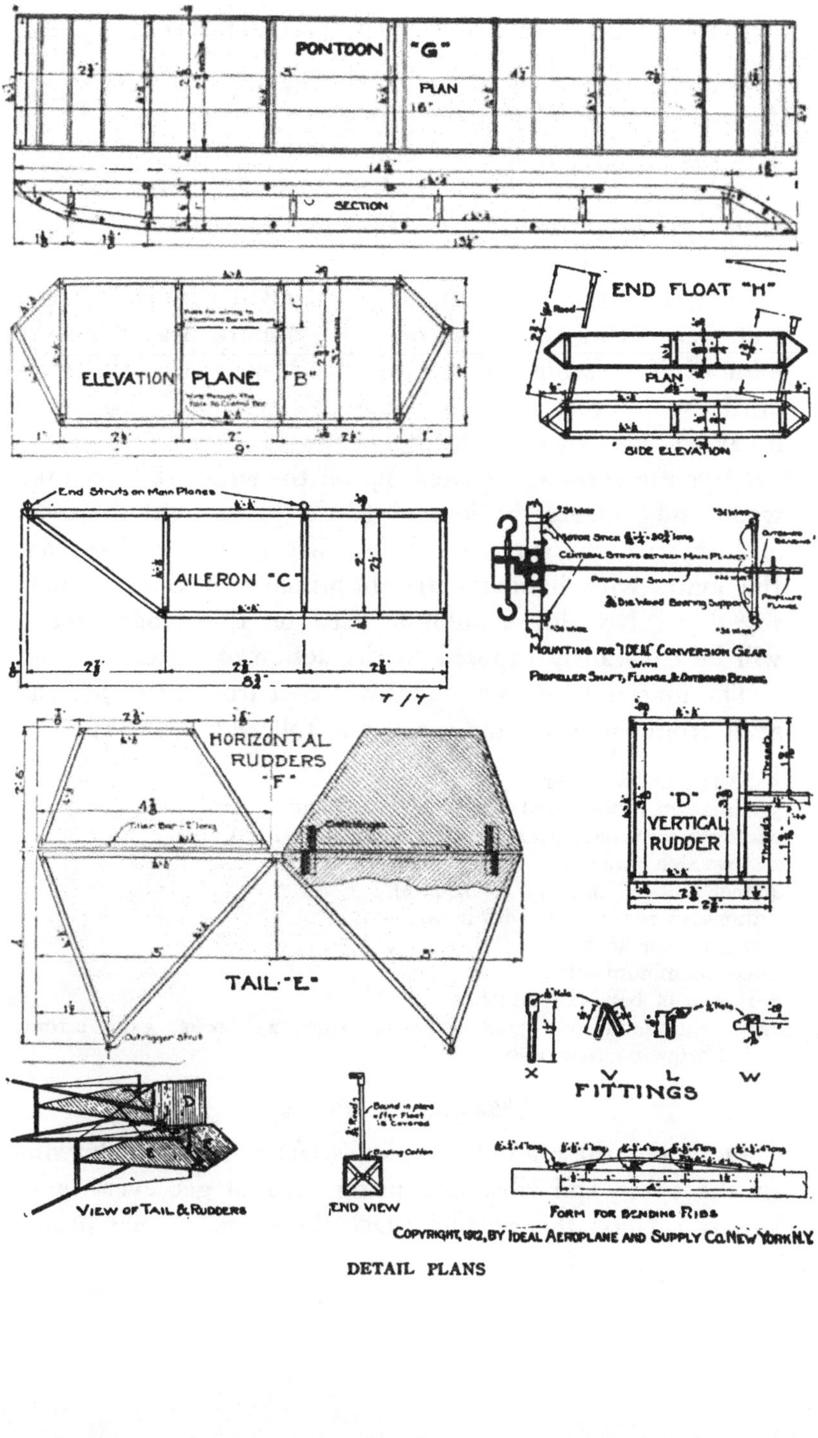


DETAIL PLANS

wood for the various parts by placing it directly over the drawings and cutting it to shape with a sharp knife drawn over the line indicating the joint. Label each piece as soon as cut with a mark that will identify it and insure its being joined in the proper place. In wiring, one turn of a wire is sufficient except on the main planes.

First twist the ends of the wires together with the fingers, and then with a small pair of pliers, pulling away from the joint while twisting the wire tight. Always twist the wires until the joint is rigid, but do *not* twist hard enough to break the wire or cut into the wood.

After the wires are twisted nip off the ends 1/8″ from the wood and squeeze the joint with the pliers until the twist is flush with the surface. Before wiring it is well to coat the joints with bamboo-varnish, which will cement them together. No glue should be used on this model, for it will be constantly exposed to the action of water.

The material for this model will cost from $7 to $8, and aside from the wood and reeds is as follows:

3 sheets bamboo-paper.
3 miniature rubber-tired wheels, 1-1/2″ diam.
Multiple conversion-gear and propeller-shaft (Ideal).
12″ wooden propeller.
40 feet rubber No. 14 (3/32″ square).
1 threaded 7-3/4″ axle with nuts.
2 rear rubber hooks.
2 feet aluminum tube.
8-oz. can of bamboo-varnish.
1/16″ aluminum rod; spool of 34-gage wire; 2 S-hooks; 3 doz. screws, and a few escutcheon-pins.

### Making the Planes

Draw out the plan of the planes (A) to full size, drawing in the spars and ribs, and using care to get every joint square. Upon this pattern place the spars in their proper position and mark on them the centers for the uprights or struts (shown as small circles in the plan), and through these drill holes with a No. 61 drill. Enlarge these holes with a 3/32″ drill, and use the utmost care not to break the spars.

Steam and bend the twenty-two ribs to form as directed for other models; or, better still, make a form by nailing parallel strips of wood to a board, as shown in detail plans. Each steamed piece is then placed on the form diagonally so as to slip both ends under the strips to hold them down. They are then worked around square to the strips, where they will remain snug and tight until dry.

While the ribs are still in the form drill holes with the No. 61 drill through each rib 1/8″ from each end.

Wire the ribs to the front spar at the points indicated in the plan, and as illustrated, making a slight notch in the front of the spar for the wire to set in, and be sure to place the ribs so they all face the same way.

Wire on the rear spar, and then the small extension ribs, not forgetting the small braces at each end. Now stretch two pieces of thread, one on each side of the plane and across the rear of the extension ribs, and tie this, and varnish at the extreme ribs; also put a drop of varnish at each intermediate rib-tip, which will be sufficient to hold the thread in place.

Cut the wood for the rib-braces—the pieces that run parallel and between the spars—and place these between the ribs, being careful to have them in a true line; these need not be fastened with wire.

When both planes have been thus constructed they should be squared up and given a thick coat of bamboo-varnish. They should then be hung up to dry thoroughly, and while drying you can make the supplementary planes.

### Supplementary Planes

These are the *elevator-plane* (B); the *tail* (E); the *horizontal rudder* (F); the *vertical rudder* (D), and the *ailerons* (C). Each of these is built in the same manner as the main planes, and all are coated with bamboo-varnish and laid upon a smooth, unpainted surface to dry, with heavy weights placed on them to keep them perfectly flat.

If placed upon a board coated with paraffin, or upon oiled paper, there is no danger of their sticking.

### Covering the Planes

When all the planes are perfectly dry they should be covered with the bamboo-paper as follows: Cut four pieces of paper an inch larger all around than the size of the planes, and with a soft brush coat each sheet evenly with bamboo-varnish. In coating the paper, or using the varnish in any way, be sure to place the object being treated on an *unpainted* surface, as the varnish *quickly destroys ordinary paint or varnish.*

As soon as a sheet of paper is varnished hang it on a cord to dry, and as some little skill is required to coat the paper quickly and evenly, you had better coat several extra sheets of paper to use in case some should prove failures.

While the varnished paper is drying lay the plane frame to be covered on a flat surface, with the hollow side up, and give the spars and ribs—for a distance of two ribs from one end—a thick coat of the bamboo-varnish. As rapidly as possible lay the cut paper over the frame and rub rapidly over the freshly varnished parts until dry. Then raise the free portion of the paper and apply the varnish to the next section of the frame, and rub down as before; and continue in this way, coating and rubbing two rib-sections at a time, until the whole plane is covered, and it can be hung up to dry.

Treat the other planes in the same way, and then cover the tops of the planes by following the same method; but in this case only the spars and ends of ribs need to be attached to the paper. In rubbing down the top paper be careful not to press so hard as to distort or flatten the ribs.

When the planes are covered on both sides and are dry, trim off the excess paper and finish smooth with a fine file, at the same time rounding the front edges of the spars slightly.

If you find that any portion of the paper has not adhered, raise it carefully, force a little varnish underneath, and rub it down. When the edges are filed smooth varnish them thoroughly.

At the rear of the planes the upper and lower coverings are fastened together, with the varnish between them, and at the rear ends of the small extension ribs, the thread already mentioned being between the two coverings, thus forming a flexible "trailing-edge." The other planes are all covered in the same way, but are kept weighted down flat until dry and ready to use.

When the main planes are dry puncture the paper where it covers the drilled holes with the 3/32" drill. If by any chance a tear or hole is made in the plane-coverings a piece of the bamboo-paper fastened over it with bamboo-varnish, and afterward coated with the varnish, will make a perfect patch.

### To Assemble the Main Planes

Cut a piece of 3/32" diameter hardwood into 16 pieces, each exactly 5-1/2" long, and proceed to assemble as follows:

Lay one plane, hollow side down, on a smooth surface and fill four holes at each end with bamboo-varnish. Press the uprights firmly but rapidly into these with a twisting motion. Place the upper plane bottom side up and fill the corresponding holes with varnish, and then quickly place it over the uprights and set each into its proper hole. Turn the planes upside down—the upper one on the table or other surface—and drive the uprights flush with the outer surfaces of the planes by gently tapping with a light hammer over each upright along the spar.

Examine the assembled planes and line them up true, twisting slightly this way and that until the two planes are parallel; and support the structure in this position until the varnish dries hard.

After it is dry it will retain its shape and will be flexible enough to allow you to insert and cement the remaining uprights one at a time, but care must be used not to bend the spars any more than absolutely necessary when placing the last uprights in position.

Next cut the 3/32″ hardwood support for the outer rear propeller-shaft bearing, and wire it between the rear-center struts in the position shown. When this is done give all a coat of varnish, and the plane-cell is complete.

The next step is to mount the ailerons in the cell by drilling two small holes at the points next to the struts and wiring the ailerons to the struts. In this model the ailerons are merely for appearance, and if omitted the model will fly even better than with them.

The horizontal rudders (F) are hinged to the tail (E) with pieces of cloth, as shown. It is best to paste the pieces on the tail with varnish, and when the varnish is hard cross the free ends and attach them to the rudders.

A small hole is then drilled through the longest side of

the horizontal rudder frames, and a wooden toothpick fastened in the hole with varnish for a tiller-bar, as plainly shown in the plan.

### The Outriggers

From the 3/32″ diameter wood cut four pieces each 12 inches long, and at one end of each drill a small hole, and then file off the end to a bevel.

Mark on the tail where these outrigger-bars are to come and drill two holes, on each side, 3/32″ apart. Place a pair of the bars between these holes, one above and one below, and wire through the holes, and do the same on the other side.

Cut two pieces of the same 3/32″ round wood, each 1–3/4″ long, and wire these to the outrigger-bars so they just touch the points of the tail.

Wire the points of the tail to these short struts a little nearer the upper bars than the lower ones. True the joints up, varnish them, and hang up to dry.

Drill one hole on the top and another on the bottom of that side of the vertical rudder (D) that is toward the main planes. Pass thread through these holes and tie off to the outrigger-bars at the struts, allowing the opening in the rudder to rest against the tail-bar. Tie another thread from the tiller-bar to the outriggers to keep the rudder in any desired position, and do the same with the tiller-bars of the horizontal rudders (F).

Make four outrigger fittings (X) by cutting off pieces of aluminum tube each 1–3/8″ long and flattening one end for 1/4″ by squeezing with a pair of pliers or by a light tap with a hammer.

The aluminum tube may be easily cut without danger of

bending it by rolling it back and forth under a sharp knife.

Drill a 1/16″ hole in the center of the flat end of each fitting, and fasten the fittings on the rear spars of the main planes by screws, placing them on the top of the lower and the bottom of the upper plane-beams. Into these fittings insert the free ends of the outrigger-bars by scraping down the bars until they slip tightly in the fittings for about 1/4″ inch.

### The Motor

Make the motor-stick from 1/4″ x 5/16″ stock, and at the center screw the "multiple conversion gear" and shaft, being sure that the shaft is free and revolves readily.

At each end of the stick drill a small hole, and place the end-hooks in these, screw on the washer and nut tight against the wood, and then mount the motor.

Place the stick in position, as shown in the general plan, allowing the end of the shaft to rest upon the bearing support, and wire the stick to the central front struts. The ends may then be loosely tied to the end struts, so that the stick may be free to twist without distorting the planes.

Next fasten the outboard bearing to the 3/32″ support already provided between the rear center struts, and coat it with varnish, being sure that no varnish remains on the shaft, and also that the shaft and gears run freely and smoothly.

The propeller should then be drilled and mounted on the shaft-flange with the hollow sides of the blades *away* from the planes, and the hub should be fastened to the flange with fine brads or escutcheon-pins.

String the rubber loosely between the hooks in the gear and the S-hooks, and hook the latter onto the end-hooks of the motor-stick. The rubber should be evenly divided into two portions and each part wound upon its own side of the stick.

### The Pontoon and Floats

The construction of the pontoon will require more time and skill than any other one part; and, while a description may help some, the best idea of its construction may be obtained by carefully studying out the plans and dimensions shown in the illustrations.

Notice that the curved portions of the sides are made of 1/4" pieces joined at a miter, as shown by the light lines outside of the curve. Wire the long and short lengths forming the top and bottom, and then join in the small cross-pieces.

When both sides are made coat them with bamboo varnish, and after a few moments place them under a weight on a smooth surface to dry flat. When dry trim away the corners until the curve is obtained.

Cut the cross-braces and wire to the sides, and note that those on the bottom are kept 1/16" from the bottom edge.

After all are joined in their correct places give them a coat of varnish and place the pontoon on a smooth board until dry, taking care that the braces are perfectly square with the sides.

The two plane-end floats are constructed in exactly the same way, with the ends drawn together in a triangular point, and the whole varnished.

Cover the sides of the pontoon by coating with bamboo varnish and paper as follows: Cut a strip of bamboo-paper 33 inches long and 3-1/14" wide, and draw a 1/4" border along each side. Bend up the paper at these lines and try it on the pontoon to see if it fits. Cut a few V-shaped notches in the borders at the curved parts of the frame so they can be lapped smoothly in place. Coat the edges of the frames and the parts of paper over the 1/4" wood with a heavy coat of bamboo-varnish and fasten the paper quickly in place, starting at the top of the frames at a cross-brace. Varnish only as much of the frame as can be covered before the varnish dries, and be sure that the paper is glued together at the lap. Touch up any loose spots; and when dry give the entire pontoon three coats of the varnish, letting each coat dry thoroughly before applying the next, and being careful to get each coat as smooth and uniform as possible. Weight the pontoon down to keep it true while drying, but leave enough of it exposed to let it dry quickly.

The small plane-floats are covered and varnished in the same way, except that a single piece of paper is glued right around the frame and is fitted over the ends by cutting and lapping.

Cut two pieces of reed for each float a little longer than required, and drill a hole at one end of each. Bind these to the plane-floats as shown, using the drilled holes for binding through, and then coat the joints with varnish, and be sure they retain their exact position while drying.

When dry cut off the reed so that when the aluminum fitting is placed on the end the total length will be that shown in the cut.

Make the L-fitting by cutting off a piece of aluminum tube 3/4" long, and with a pair of pliers squeeze the tube flat for a distance of 1/4" from one end. Bend the flat part at right angles to the round part, being careful to bend it slowly. In the center of the flat portion drill a 1/16" hole for the screw.

These fittings (18 are needed) are worked on to the end of the reeds used as supports, as shown in plan and general view, by twisting back and forth, but great care must be used not to twist the reed so as to weaken it. Also be sure and get the fittings so each turns the proper way, which can be done as follows:

Make a plan of the pontoon on a board, as already suggested, and draw in their proper location the forward cross-brace and the three 1/8″ x 1/8″ cross-braces, all on top of the pontoon. Place the L-fittings on the end of a piece of reed cut for the braces, and screw them to the board in the proper position, as indicated by the diagram on the board. Cut the supporting-uprights to their exact lengths and place the upper fittings in position.

Cut a piece of aluminum rod 3–3/8″ long, and slowly bend 1/2″ down on each end and bind this rod to the uprights; this forms the axle of rotation for the elevating-plane.

Next cut two pieces of reed 10 inches long, and drill a small hole at each end. Wire one end to the elevating-plane uprights near the top, and bring the two pieces together over the first 1/8″ square cross-pieces, as shown on the board diagram. Clamp down by screwing over them the aluminum clamp (W), being careful not to pull the uprights away from their vertical position.

Wire the other ends of the reed to the next set of uprights, keeping them vertical. The elevating-plane control-bar is also of reed, and is fastened between the reeds just mentioned and in front of the clamp (W). When all are in place touch each joint with bamboo-varnish, and later coat the whole chassis with varnish also.

When dry carefully remove the assembled chassis from the board and screw it on the pontoon, being careful to first drill small holes and see that the fittings are screwed to their proper cross-braces.

The fitting (W) referred to is made from a piece of aluminum tube 5/16″ long flattened out with a hammer, and with 1/16″ bent down with pliers on each side; a 1/16″ hole is made in the center for the screw.

The elevating-plane may now be wired to the aluminum bar, and the control-bar fastened through the proper holes, as shown in the cuts.

Screw the uprights of the pontoons and small floats into the spars of the main planes (being sure to drill holes first), and while screwing in place *always* support the spars from beneath with the hand directly under the screw, so as to be sure not to break the slender wood. Also be very careful not to let the screw-driver slip and tear the plane covering. To secure greater stiffness the pontoon uprights should be cross-braced with thread or very light reed diagonal braces. This will complete the hydroaeroplane, and it may be flown from the water as it is.

It is advisable, however, to build a land running-gear and chassis also, so that you can fly the model from land when water is not at hand, and you should proceed as follows to build a

### Land-Chassis

First make two V-shaped fittings (V, plan of fittings) from pieces of aluminum tube about one inch long bent sharply at the center with the hands, bending it so the ends are separated at the angle shown in the cut. Then with pliers flatten the portion at the bend and drill a 1/16″ hole through it to fit the axle.

Moisten a piece of reed and hold it a few seconds over a flame, and then bend it to the form of the front fork shown in view of land-chassis. Trim the ends to equal length, and fasten each end into one arm of a V-fitting. Cut the 1/8″

x 1/8″ keel-piece and drill a hole at one end, and wire on the reed fork just made.

Varnish the joint and let it dry, and when hard pass the axle of aluminum rod through the holes in the V-fittings—with the wheel between the two—and flatten the projecting ends with pliers to prevent the axle from slipping out.

Cut a piece of reed about nine inches long, bend it sharply at the center by moistening and holding over a flame, and wire it at the bend to the keel-stick in the position shown in view of chassis. Cut the ends to the right size and slip L-fittings in place. This forms the "front-strut" to the chassis for the running-gear. Next place the two front reed braces into the empty arms of the V-fittings at the axle, and wire the free ends to the front-strut near the top.

Cut the rear uprights and fit on L-fittings and bind the other ends to the steel axle, leaving 3/4″ of the axle free at each end for the wheels.

Bind the center of the axle to the keel-stick in the position shown in view of chassis, cutting a slight groove in the top of the stick for the axle to rest in. Run the diagonal reed braces as shown in plan; one bent piece from the top of the rear uprights to the axle where it joins the stick, and two pieces from the top of the front-strut to the foot of the rear uprights.

True up the chassis and varnish it, putting an additional drop of varnish at each joint. When dry place the rear wheels on the axle-ends and fasten with nuts.

The land-chassis and running-gear is now complete and may be screwed onto the plane-spars in place of the pontoon; but the rear uprights of the former need extra holes drilled in the rear plane-spars. These holes must be varnished *inside* to prevent any water from entering the wood, and when using the machine as a hydroaeroplane it is a good plan to fill the holes with varnish, which may easily be removed when you wish to transform the model to a land-machine.

### Flying the Model

The motor-rubbers may be wound up either by turning the propeller by hand or by using a regular winder on each half. By the former method you can wind and fly the machine by yourself, but as a smaller number of turns are obtained in this way, and more time is required, it is always better to use a winder and have a friend to help, when possible. In case the motor is to be wound by hand the S-hooks at the ends of the rubbers may be omitted.

In winding by hand hold the gear at the motor-stick with one hand and turn the propeller backward with the other, giving it about 250 turns.

Be sure that the rubbers are always wound in the right direction; viewing the machine from the back the propeller should turn like the hands of a clock, from left to right.

After winding keep hold of the propeller with your hand and note if all parts of the plane are true and parallel. If the model is equipped with the pontoon, merely let it rest gently on the water, give it a slight push, and release the propeller. The same method may be used on land with the running-gear attached, or it may be flown from a smooth, upward-tilted surface, as directed for other models.

Still another method is to hold the model level and pointed slightly upward in your hand, and gently cast it forward, lowering the hand very quickly to avoid catching it in the tail.

A fairly calm day should be selected for flying the machine, and after a little practice you will learn just how to set the rudders and elevators to obtain the best results in all sorts of weather.

To rise from land or water quickly, the horizontal rudders should be turned up slightly at the rear, and when the pontoon is used the elevator must have its front edge a little higher than its rear edge.

The vertical rudder is used to turn the model to right or left, and the machine may be made to travel in a curve by this rudder, or it may be set to one side just enough to overcome any side-drift or irregular course due to wind or uneven balance or unequal construction of some part.

### The Miniature Flying-Boat

Even more interesting than the model Curtiss hydroaeroplane is the model Curtiss flying-boat perfected and placed on the market by the Ideal Aeroplane and Supply Company.

This beautiful little aircraft is an exact counterpart of the real and original "Flying-boat," and by the aid of the plans and directions furnished by the Ideal Company any boy may readily construct this most recent model.

The dainty little craft will fly equally well from the hand, on water, or from land; but in the latter case it must be launched from a "land-chassis" which consists of three rubber-tired wheels on a light wire frame. In use the chassis remains behind when the flying-boat rises.

The model is thoroughly waterproof and cannot be injured by striking the water or being submerged entirely.

The construction of these two water flying-models will keep you and your friends interested for a long time, and I have no doubt that you will soon commence building other types of hydroaeroplanes, or may even attempt to construct a real man-carrying flying-boat.

## HOW TO BUILD A HYDROAEROPLANE

The knowledge you have gained by building models will serve you well if you undertake this, and by merely using care, common sense, and plenty of time you can succeed just as well at the full-sized machine as with the little models.

# Part VI

# USES OF THE AEROPLANE

# Chapter XVII

## AEROPLANES IN PEACE AND WAR

WHEN man first conquered the air and aeroplanes became an actual fact most wonderful predictions were made as to their future usefulness. People thought that in a few years regular aerial vehicles would carry passengers, freight, and mails from place to place on schedule time, like railway-trains or steamships.

Papers and periodicals contained highly interesting and graphic accounts of future wars to be fought with airships, and, in fact, the world went aeroplane-mad.

As time passed on and the real capabilities and scope of aeroplanes became better known people realized that, useful and wonderful as they are, aeroplanes have limitations, and except in certain lines are of little general use commercially.

That well-built and well-handled aeroplanes can carry mail, passengers, and freight is indisputable, for this has been accomplished time and again.

So uncertain is the air, and so dependent are the best aeroplanes upon weather conditions, that trips cannot be made with certainty or on schedule, even if the engine does not fail and no accident occurs.

For transporting special messages or despatches rapidly there is no doubt that the aeroplane will be of value both in peace and war; but that it will ever supplant railways and steamboats as a regular method of transportation is very doubtful.

Passengers can be and are safely carried by aeroplanes, and as many as seven passengers have been carried at one time.

As a rule these trips are made merely for the novelty of the experience, but on several occasions doctors have found aeroplanes of the greatest value when compelled to travel a long distance at utmost speed.

As aeroplanes readily travel from 60 to 75 miles an hour, and can travel in a straight line from place to place, they can transport a passenger or mail with the utmost despatch when required.

In war-time aeroplanes also play quite an important part, and, whereas their true value in warfare was questioned until really tested, they have now proved to be of great use in actual hostilities.

In Tripoli, in the Balkans, and in Mexico aeroplanes have played an important part in war and have proved of very great value.

As scouts, in learning the position of fortifications and guns, and in furnishing reports of surrounding country, as well as in carrying despatches from one officer to another, aeroplanes are wonderfully useful.

To be sure, an aeroplane is now and then struck with a shell and brought to earth, but even if the operator is killed the loss of a single machine and operator in this way is of no great importance as compared with the service rendered.

Moreover, a falling aeroplane, or even a falling aviator, may kill or disable a number of the enemy, and in a way a dead aviator or a crippled machine is far more dangerous to life and limb than one flying overhead.

Aeroplanes are very cheap as compared to other warlike apparatus, and the same amount spent in aeroplanes that is spent on a single war-ship would probably in the end bring far greater results in time of war.

Not only can aeroplanes rise to great heights and still observe the enemy and his camps and forts, but excellent photographs may be taken from the aircraft.

Such pictures show the country beneath even better than a map, and with modern photographic appliances and inventions the photographs may be taken, the negatives developed, and the pictures examined within a few minutes; in fact, negatives are often dropped from the aeroplanes to earth by means of small parachutes.

Many splendid biograph pictures have been taken from aeroplanes, and no doubt aeroplane photography may at some time become a popular fad.

Numerous army and naval officers have made excellent map sketches from aeroplanes in flight. The first aerial map made by an army officer in a long-distance flight (Fig. 1) was drawn by Lieutenant Sherman, U.S.A., on the return non-stop cross-country flight between Texas City and San Antonio, Texas, on March 28 and 31, 1913.

The total distance covered in this flight was 480 miles, the machine used being a Burgess tractor biplane equipped with a Renault motor. It was planned to make this flight by compass, but the air was very hazy and rough, and after striking the Sante Fê the railway was followed to San Antonio.

A portion of this country is flat and treeless, but from Eagle Lake to San Antonio the country is rolling and covered with forests interspersed with cultivated tracts.

The air was so rough and the temperature so high that great difficulty was encountered in handling the plane, which on one occasion dropped to 600 feet, and it was fre-

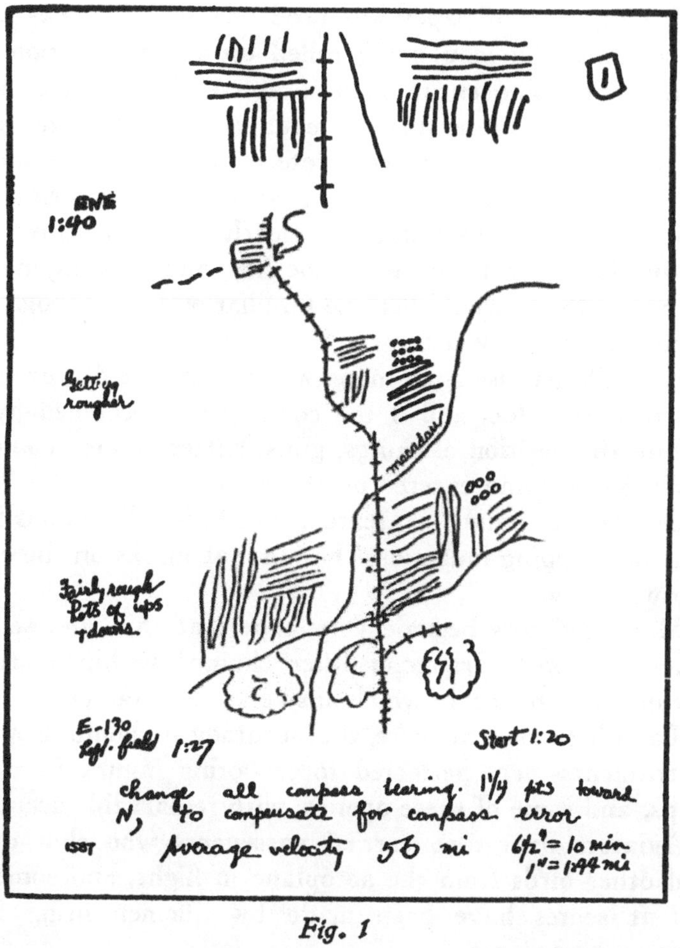


*Fig. 1*

quently necessary to dive for 50 to 100 feet to regain equilibrium.

Lieutenant Sherman carried a cavalry sketching-case, and only attempted a rough sketch. The sketching-board was held parallel to the fuselage and the compass-bearing noted and a time-scale used.

A long strip of paper was used and the map was made in sections, the map being rolled up as each section was completed. The entire map is about eighteen feet long, each section representing the country covered in ten minutes. The map is very complete in detail, and shows the railroad, wagon-roads, streams, woods, hills, prairies, and other features of the country so clearly that any army could readily locate each and every locality, and by studying the map the officers would become familiar with the topography of the country covered.

For military use such maps would undoubtedly prove of the utmost value, and if the country was occupied by an enemy the position of camps, guns, batteries, etc., could be easily located and entered on the map.

One of the commonest feats in exhibition flights has been that of dropping imitation "bombs" at marks on the earth below.

Many aviators became very expert at this, and seldom failed to drop an orange or other "bomb" within the rough outline of a steamship which answered for a target.

To still further increase the accuracy of bomb-dropping instruments were perfected for shooting bombs from airships, and some of these operate with remarkable accuracy.

Aviators have also carried passengers who shot ducks and other birds from the aeroplane in flight, and some excellent scores have been made by riflemen firing from moving aeroplanes.

As a whole, however, aeroplanes are mainly of value for sporting or exhibition purposes, and their inherent danger has hitherto prevented many ardent sportsmen from adopting them.

The advent of the hydroaeroplane and the flying-boat has made flying as safe as motor-boating, and probably within a year or two we will see almost as many flying-yachts in use on our bays, rivers, and lakes as we nowadays see real yachts and launches.

### The Sensations of Flight

Many people are very curious to know just what the sensations of flying are, and many are afraid to go up fearing that they will be dizzy or have an inclination to jump out. I have never known of a person feeling dizzy or giddy in an aeroplane, and, although the passenger is seated on a frail seat with nothing beneath but the air, and can look directly down to the earth a thousand feet below, yet there is a feeling of safety and security that is entirely absent when looking from a high building or a mountain-top.

Each and every person has slightly different sensations when making his or her first flight. One man described his sensations as similar to riding on a fluttering flag. Another compared it to riding on a huge feather, while others state that it seems like a dream in which the body is suddenly bereft of all sensation of weight or bulk.

The author's eighteen-year-old daughter made a long and high flight with Harry Brown in a Wright machine and stated that the first wild rush across the field, before rising, was the most thrilling and exciting part of the trip, and that as the machine rose the earth seemed to drop suddenly from beneath. After the biplane once left the earth and began to rise the ascent appeared to be very gradual, and the machine seemed to poise and remain motionless, swaying gently in the wind, although the apparent gale was really the motion of the rapidly traveling machine through the air. So safe and steady felt the biplane that the passenger released her grasp on the stanchions and used her hands to arrange her hair and ribbons.

The machine was then turned and "banked" sharply, and at this moment the first real flying-sensation was felt. As the aeroplane swerved and tore at sixty miles down the wind the Long Island landscape unrolled like a vast panorama beneath, with all details queerly foreshortened and distorted.

The roads seemed erratic and at varying angles, the fields irregularly shaped, with the shrubs and bushes looking like dark-green clover on a lawn. At this time the aeroplane was about fifteen hundred feet in the air, and, while trees were readily recognized, the houses, wagons, horses, automobiles, and people looked like exquisitely formed toys, and a little red runabout appeared like a tiny potato-bug. There was no sensation of height or distance; everything below seemed merely dwarfed in size; and no feeling of exhilaration, such as is felt when speeding on land or water, was experienced. Instead there was a feeling of aloofness, a sort of mental numbness such as one feels in certain dreams.

I doubt if any two people ever have exactly the same feelings while flying, but the fact that hundreds of people of all ages and temperaments are carried through the air each season proves the safety and ease of taking an aerial voyage with a competent pilot.

### Are Aeroplanes Dangerous? Some Facts and Figures

Many persons insist that aviation is a most hazardous profession, and that the man who attempts a flight takes his life in his hands.

In reality facts, borne out by indisputable evidence and

records, prove that conservative aeroplaning is as safe as automobile riding, and is safer than football.

In the early days of aviation each flight was heralded throughout the world, and every successful and every disastrous flight was recorded.

To-day flights are so common, so numerous, and such an every-day matter that only notable record-breaking flights or serious accidents are published in periodicals and newspapers.

This draws the attention of the public only to accidents, and the average man does not and cannot realize that the majority of flights are successful, and are so numerous that they cannot be recorded.

Few can believe that in the four years since the first aeroplane made public flights the number of aviators has increased 1,200 times!

How few of us realize that during 1912 no less than 12,000,000 miles were traveled in aerolpanes, or that 6,000 men were actively engaged in flying as recognized pilots or aviators!

In 1903 only 1,000 miles were flown, and but five trained aviators were known. One year later the mileage increased to 28,000, and 50 aviators had learned to fly. By 1910 no less than 500 airmen were flying, and aeroplanes traversed 600,000 miles through the air.

The figures increased rapidly, and by leaps and bounds, and at the close of 1911 1,500 aviators flew 2,300,000 miles, which during 1912 increased to 12,000,000 miles flown, or a distance equal to 24 round trips to the moon, while the army of airmen had grown to 6,000.

These figures illustrate how rapidly aviation has grown, and if we look at the records of accidents and deaths we will find that these have *decreased* even more rapidly in proportion.

In 1908 one out of five aviators was killed; in 1909 only four out of fifty met death by flying; in 1910 the number of deaths was 30 out of 500; in 1911 only 77 of the 1,500 aviators were killed, and in 1912 the proportion was 152 deaths to 6,000 men.

The total number of fatalities since aeroplanes first flew until June 1, 1913, was 264, and as aeroplanes have altogether flown 14,929,000 miles, there has been but one fatal accident to each 56,549 miles flown.

In other words, safety in aviation has increased over 1,000 per cent. in a single year (1911), and in the past year (1912) has jumped to a 4,000-per-cent. increase.

These figures apply only to professional aviators, and take no account of the vast number of passengers carried, and of whom but twenty-three have been killed, including military officers who are classed as passengers. Just how many real passengers have been carried cannot be determined; but many aviators have carried thirty-five passengers in a day, and have made over three hundred passenger-carrying trips in a season without a single accident.

It is very seldom indeed that an accident occurs when reasonable care is taken and the aviator uses common sense and does not attempt spectacular feats.

The great majority of accidents—fatal and otherwise—have been due to gross carelessness or performing "stunts." Men like Lincoln Beachy, who appears to bear a charmed life, perform feats of a most daring and death-defying character, such as flying with both hands free, flying over crowds, flying through the gorge of Niagara, etc.

Many other airmen have attempted similar feats with disastrous results, and the unknowing at once set down such accidents as proving the dangers of aviation, whereas the unfortunate victims would probably have killed themselves just as quickly were they operating automobiles or other high-powered machines.

Carelessness, daredeviltry, and recklessness have no place in aviation, and it is seldom indeed that a cool-headed, cautious, and conservative aviator meets with serious accidents.

The Wright brothers made hundreds of flights, many of them in untried and experimental machines; Capt. Baldwin has grown gray in aviation; Glenn Curtiss is a veteran of the air; and Cecil Peoli, a mere boy, has made over four hundred flights in a season, and no death or serious accident has ever marred the career of any one of these famous aviators.

Aeroplanes are so new and have been developed and perfected so rapidly that a few deaths and accidents were unavoidable, as in any new art, and the great wonder is that more fatalities have not resulted from flying.

No one can foresee the future of aviation, and each year, month, and day brings forth new inventions, new discoveries, and new devices to make human flight safer, more certain, and more exact, and we may confidently look for undreamed-of progress in the future.

No art or discovery of mankind has ever made such rapid progress as aviation, and aeroplanes—still in their infancy—have been perfected more rapidly than any other mechanical device ever built by human hands.

## International Aeroplane Records

**Revised to May, 1913**

| | | 1-Man | 2-Man | 3-Man | 4-Man | 5-Man |
|---|---|---|---|---|---|---|
| Duration | | 13:17:57.2 | 4:34:00.0 | 2:41:00.0 | 1:35:00.0 | 1:18:00.0 |
| Distance | | 1010.9 kil. | 401.50 kil. | 112 kil. | 110 kil. | 25.74 kil. |
| Altitude | | **5610 m. | 4360 m. | 3580 m. | 1120 m. | ***596 m. |
| Greatest Speed | | ‡174.10 k.p.h. | 135.952 k p h. | 102.855 k.p.h. | 106.029 k.p.h. | 87.251 k.p.h. |
| Climbing Speed | 500 m. | ‡3:35.00 | ...... | ...... | ...... | ..... |
| Climbing Speed | 1000 m. | 4:56.50 | ‡9:00.0 | ...... | ...... | ..... |
| Speed | 5 kil. | *‡1:43.38 | 2:58.0 | 2:52.0 | 3:48.0 | 3:34.0 |
| " | 10 kil. | *‡3:27.87 | 4:24.8 | 5:45.0 | 6:16.6 | 7:08.0 |
| " | 20 kil. | *‡6:55.95 | 8:51.0 | 11:59.4 | 12:03.0 | 14:00.6 |
| " | 30 kil. | *‡10:32.51 | 13:18.6 | 17:52.6 | 17:37.0 | |
| | 40 kil. | *‡14:03.59 | 17:44.8 | 22:44.4 | 23:11.0 | |
| | 50 kil. | *‡17:34.88 | 23:13.0 | 29:37.4 | 29:47.0 | |
| | 100 kil. | *‡35:16.65 | 44:36.6 | 59:08.0 | 56:33.0 | |
| | 150 kil. | 52:52.80 | 1:07:10.0 | ...... | ...... | **6-Man** |
| | 200 kil. | 1:10:55.00 | 2:03:49.0 | ...... | ...... | Duration |
| | 250 kil. | 2:07:54.00 | 2:39:37.0 | ...... | ...... | record only, |
| | 300 kil. | 2:49:00.00 | ...... | ...... | ...... | 1:06:48.2 |
| | 350 kil. | 3:26:16.00 | ...... | ...... | ...... | |
| | 400 kil. | 3:55:27.60 | ...... | ...... | ...... | |
| | 450 kil. | 4:24:44.40 | ...... | ...... | ...... | |
| | 500 kil. | 4:54:06.20 | ...... | ...... | ...... | |
| | 600 kil. | 5:52:38.00 | ...... | ...... | ...... | |
| | 700 kil. | 9:31:01.00 | ...... | ...... | ...... | |
| | 800 kil. | 10:44:45.80 | ...... | ...... | ...... | |
| " | 900 kil. | 11:59:09.60 | ...... | ...... | ...... | |
| " | 1000 kil. | 13:01:12.00 | ...... | ...... | ...... | |
| Time | ¼ hr. | 45.664 kil | 31.020 kil. | ...... | ...... | |
| " | ½ hr. | 84.665 kil. | 66.639 kil. | ...... | ...... | |
| " | 1 hr. | 168.244 kil. | 133.469 kil. | ...... | 106.029 kil. | |
| | 2 hrs. | 234.431 kil. | 190.858 kil. | ...... | ...... | |
| | 3 hrs. | 310.281 kil. | 224.850 kil. | ...... | ...... | |
| | 4 hrs. | 410.900 kil. | ...... | | | |
| | 5 hrs. | 510.000 kil. | ...... | | | |
| | 6 hrs. | 490.000 kil. | ...... | | | |
| | 7 hrs. | 522.935 kil. | ...... | | | |
| | 8 hrs. | 585.200 kil. | ...... | | | |
| | 9 hrs. | 661.200 kil. | ...... | | | |
| | 10 hrs. | 744.800 kil. | ...... | | | |
| | 11 hrs. | 820.800 kil. | ...... | | | |
| | 12 hrs. | 904.400 kil. | ...... | | | |
| | 13 hrs. | 980.400 kil. | ...... | | | |

** This record since broken, though not yet official, by altitude of 6000 m.

‡ Made in United States.

*** Since beaten.

* Revised speed records made by Vedrines at Chicago. Slower records previously reported to F. A. I. are included in F. A. I. report as official.

Practically all aeroplane records are held by Bosch equipped motors.

## American Aeroplane Records

| | | 1-Man | 2-Man | 3-Man |
|---|---|---|---|---|
| Duration | | ‡‡6:10:35.0 | 4:22:00 | 1:54:42.6 |
| Distance | | 283.628 kil. | ...... | ...... |
| Altitude | | 3548.5 m. | 1422 m. | ...... |
| Greatest Speed | | ‡174.10 k.p.h. | 101.762 k.p.h. | 56.263 k.p h. |
| Climbing Speed | 500 m. | ‡3:35 | ...... | ...... |
| Climbing Speed | 1000 m. | ...... | ‡9:00 | ...... |
| Speed | 5 kil. | *‡1:43.38 | ..... | 6:56.4 |
| " | 10 kil. | ‡3:27.87 | 6:13.4 | |
| " | 20 kil. | ‡6:55.95 | 12:26.6 | **4-Man**** |
| | 30 kil. | *‡10:32.51 | 18:42.0 | Duration |
| | 40 kil. | *‡14:03.59 | 24:49.8 | record only, |
| | 50 kil. | *‡17:34.88 | 31:01.6 | 1:54.0 |
| | 100 kil. | *‡35:16.65 | ...... | |
| | 150 kil. | 53:04.73 | | |
| " | 200 kil. | 1:10:56.85 | ...... | |
| " | 250 kil. | 3:32:56.4 | ...... | |
| Time | ¼ hr. | 40 kil. | 24.14 kil. | |
| " | ½ hr. | 80 kil. | 36.24 kil. | |
| " | 1 hr. | 166.6 kil. | | |
| " | 2 hrs. | 141.97 kil. | | |
| " | 3 hrs. | 214.57 kil. | | |
| " | 4 hrs. | 283.628 kil. | | |
| Alighting, from Mark | | 0.445 m. | | |
| Weight-carrying | | 458 lbs. | | |

* Passed by A. C. A. as American record but not yet by F. A. I. as an international ore

‡‡ Hydroaeroplanes.

‡ World records.

** Not yet passed.

### MISCELLANEOUS WORLD RECORDS

BALLOONS
Distance—*2191 kiloms.
Duration—73 hours,
Altitude—10,800 meters.

DIRIGIBLES
Distance—807 kiloms.
Duration—7 hours, 13 min.
Altitude—3080 meters.
Speed—37.808 kils. per hour.

KITES
Altitude—‡7265 meters.

SOUNDING BALLOONS
Altitude—‡30,486 meters.
‡ Made in United States.

### MISCELLANEOUS AMERICAN RECORDS

BALLOONS
Distance—1887.6 kilometers.
Duration—48 hours, 26 minutes.
Lahm Cup—1172.9 miles.

DIRIGIBLES
Speed—31.559 kil. per hour.
Duration—2 hours, 1 minute, 50 seconds.

KITES
Altitude—*7265 meters.

SOUNDING BALLOONS
Altitude—*30,486 meters.
* World record.

# Chapter XVIII

## MISCELLANEOUS AIRCRAFT

AEROPLANES and gliders are known as "heavier-than-air machines," for they weigh more than the actual air they displace.

Quite a different class of aircraft are the lighter-than-air machines, more commonly known as balloons.

Formerly all balloons were merely huge silken or cloth bags filled with hot air or gas, which, being lighter than air, caused the whole fabric to rise from the earth.

In balloons men traveled from place to place at the will of the winds, their only control over their conveyance being ability to rise by throwing out ballast, or to descend by opening a valve which allowed a portion of the gas to escape from the balloon.

Balloons are still widely used as a means of elevating parachutes, in which men descend to earth at fairs and exhibitions, and a number of aeronauts still use balloons in races or contests in which the balloons often travel long distances and in unlooked-for directions.

### Dirigibles

Long before aeroplanes were invented balloons had been constructed which were provided with propellers operated by engines or foot-power, and which were capable of being forced through the air and steered from place to place at will.

These were known as "dirigibles," and to-day dirigibles have been developed to a very great degree (Fig. 1).

The Zeppelin and other European dirigibles carry a number of people on regular trips, and in one of these craft Mr. Wellman attempted to cross from America to Europe.

*Fig. 1*

*a*—Vertical rudder.
*b*—Horizontal rudder.

Some of these dirigibles are enormous structures 500 feet or more in length and 75 or 80 feet in diameter. These great balloons are usually composed of an elongated covering or envelope containing a number of small inflated balloons, and thus the danger of a puncture, leak, or tear affecting the whole is minimized.

The modern dirigibles are provided with powerful motors driving the propellers, and have horizontal and vertical rudders, stabilizing fins, and similar devices, some of which are very plainly shown in the illustration.

Beneath the real balloon a long car, or "nacelle," or two or three smaller cars, are suspended, and these contain the machinery, steering-apparatus, passengers, etc.

The sustaining power of these monstrous airships is very great, and a number of passengers may be carried at one time. The Zeppelin dirigibles make regular passenger-carrying trips, but the greatest difficulty is often encountered in landing.

Several of the largest dirigibles have been wrecked by being blown against trees, buildings, etc., while others have been burned by the gas in the balloons catching fire.

Some dirigibles are quite fast, reaching a speed of fifty miles an hour, but the majority are very slow as compared with aeroplanes.

They are very costly, bulky, and unwieldy, and cannot be safely handled in strong winds.

To protect and house them while idle, huge steel and iron sheds are necessary, and a small army of men is required to get them in and out of these sheds.

Most aeronauts have abandoned the balloon types in favor of aeroplanes, but many European countries still adhere to them for military and commercial purposes.

One of the greatest objections to balloons of any type is the danger of explosions and fire. As they are filled with highly inflammable gas, a slight spark or a fire, or even a crossed or short-circuited wire from the batteries or magneto, will often result in death or injuries to the passengers.

This was the fate of the dirigible in which Mr. Vanniman hoped to cross the Atlantic Ocean, and in the explosion and fire that occurred the maker and his fellows perished miserably, with no chance of escape.

Vanniman's dirigible, the *Akron*, was one of the most

complete, luxurious, and carefully planned and constructed dirigibles ever made, and its untimely fate was not due to any inherent fault of the machine, but to unexpected and unavoidable troubles which menace every aircraft carrying inflammable gas for buoyancy.

To overcome the loss of gas by expansion under heat the gas-bag of the *Akron* was made of heavier material than ever before used in dirigible construction, thus permitting further compression of the hydrogen gas by interior air-chambers in the form of smaller balloons which could be filled or emptied at will.

By pumping these internal balloonettes full of air the hydrogen about them was compressed, thus decreasing the buoyancy of the dirigible and its lifting power, whereas by exhausting the air the buoyancy was increased. This method was designed to overcome the loss or increase of bouyancy due to expansion and contraction of the gas by sunshine or cool night air. In addition the machine was provided with changeable water ballast which was scooped from the sea by a special device known as a "hydroleveter," and consisting of self-filling tanks which were wound up by motor into the car on a cable.

The cloth or fabric used as covering was a specially designed material with steel piano-wire as the woof and cotton as the warp. In use this fabric was laid over a form in strips running lengthwise with the steel wires in unbroken lengths one inch apart, and the overlapping seams cemented and vulcanized as fast as applied.

About these strips other sections were wound or rolled, cemented, and vulcanized to the lower layer, thus giving a double cross-weave of wires. On the outside the shell was further coated with rubber painted and varnished. The envelope was then inflated with air and the inside coated with a lining of a secret jelly-like substance which prevented any possible escape of gas.

The total length of this enormously expensive dirigible gas-bag was 525 feet, with a diameter of 52 feet.

Below the bag was the car, 200 feet long, with an upper deck, cabins, dining-saloons, smoking-rooms, kitchens, promenade, etc. (Fig. 2).

The engines were placed at both front and rear, and the craft was provided with every appliance for controlling and navigating the great balloon. Notwithstanding the care used in its construction the entire fabric was destroyed in a brief moment by a spark igniting the gas, and the work of many months, as well as the lives of the inventor and his crew, were snuffed out instantly.

Although Europeans have devoted far more time to the perfection of dirigibles than Americans, yet recently several American aviators have turned to dirigibles.

A machine which promises to be very successful has been constructed by Mr. Roy Knabenshue and Mr. Walter Brookins, both well-known and famous aeroplane enthusiasts and aviators.

This machine is but 150 feet in length and 30 feet in diameter. It contains 76,000 cubic feet of gas, and has a lifting-power of 4,940 pounds. The balloon and balloonette weigh 1,120 pounds, and with car complete with water, 25 gallons of gasolene, and appliances, 1,378 pounds, leaving a net lift of 2,392 pounds.

The machine is driven by a 35-horse-power Hansen motor, with two propellers, which drive the dirigible at 30 miles an hour. The car, which is suspended by Roebling steel cable 1/8″ in diameter, is 112 feet long and 7 feet below the bag.

The car is equipped with aeroplanes in front and rear,

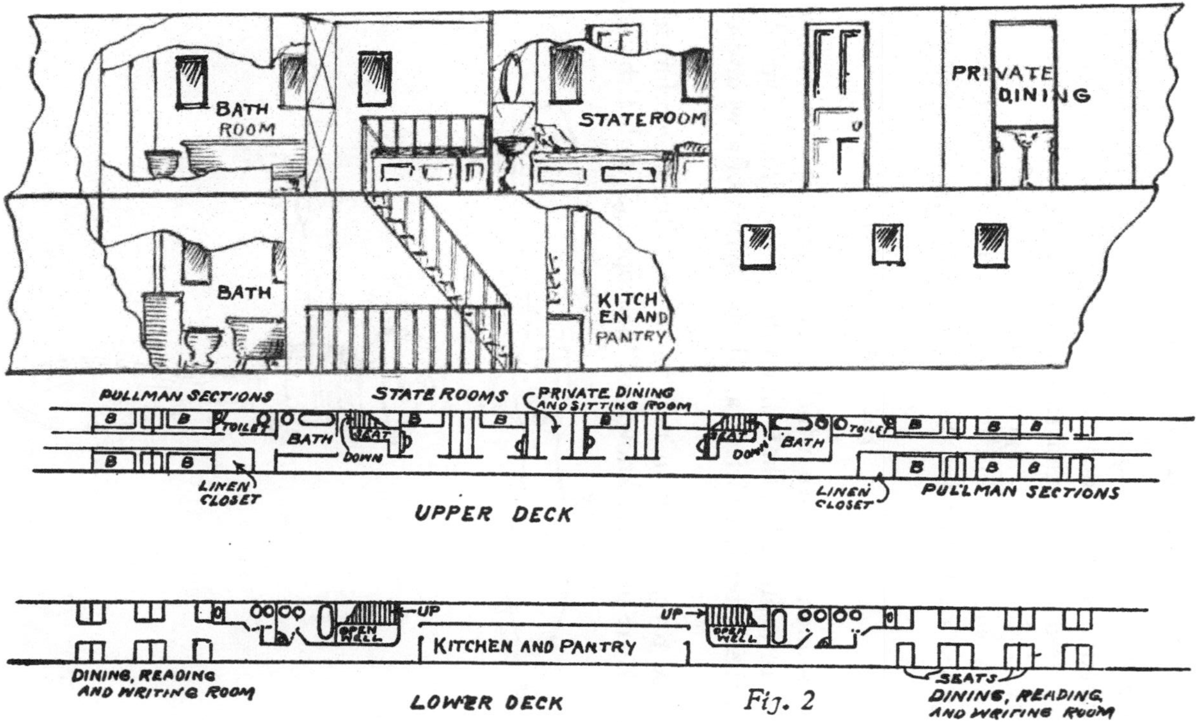


Fig. 2

PLAN VIEWS OF THE UPPER AND LOWER DECKS OF VANNIMAN'S NEW DIRIGIBLE each containing 120 square feet of surface, and these, in conjunction with rudders 6 by 10 feet, enable the balloon to be tilted, steered, and turned with perfect ease. In its recent trials at Los Angeles this new American dirigible was very successful, and its inventors are confident that it will prove a safe and reliable means of navigating the air.

### Helicopters

A common toy with which nearly every boy is familiar is the tiny propeller that flies into the air from a spindle when the latter is whirled rapidly by means of a wound-up string.

These toys, known in my youth as "Flying Dutchmen" (Fig. 3), are properly known as "helicopters," and, although

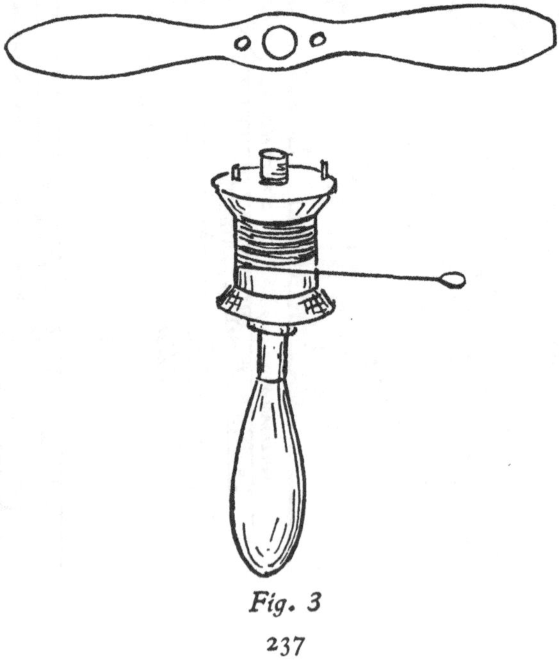

*Fig. 3*

they fly splendidly as toys, yet in large man-carrying-sized machines they never succeed.

Various men have spent time, money, and thought in designing and building helicopter machines, but the greatest altitude ever attained has been but a few inches.

While the principle of the helicopter—which is a horizontal propeller designed to lift the machine directly into the air—is correct in theory, yet in practice the mechanical problems to be overcome have hitherto proved insurmountable.

### Ornithopters

Still another class of flying-machines on which vast sums of money and a great deal of valuable time have been wasted are the so-called "ornithopters," or birdlike flying-machines, in which two or more wings beat the air like the flapping wings of a bird or insect.

Like the helicopters, these devices prove successful only in model or toy form, and all full-sized ornithopters have proved failures, and with our present facilities and knowledge of mechanics we cannot hope to produce ornithopters or helicopters capable of lifting a man in flight.

### Freak Aircraft

In addition to balloons, helicopters, and ornithopters, many freak aircraft have been designed and tried.

Some of these are combinations of helicopters and ornithopters, while others combine the peculiarities of these devices with some aeroplane features.

Still others are designed to fly by paddle-wheels; others by a wavelike motion of strips of fabric, and still others by man-power. The latter, known as "aviettes," have actually succeeded in "jumping" from the earth and sailing short distances; but all others have proved most dismal failures.

That man can propel a light, properly constructed plane with sufficient speed and power to actually lift his own weight may yet be proved, and in France a great deal of interest has been shown in the prize competitions, and great efforts have been made to produce a practical aviette.

Even the smallest of power-driven aeroplanes requires several horse-power to raise it from the ground and force it through the air in flight, and for man to develop sufficient energy to equal the weakest aeroplane motor is an impossible task with any known mechanical devices.

## Box and Tetrahedral Kites

Much of the knowledge gained in regard to the action of air currents on planes and surfaces was due to experiments with various forms of kites. The Hargrave box-kites (Fig. 4) and the tetrahedral kites of Alexander Graham Bell, as well as the man-carrying kites of S. F. Cody, were all remarkably efficient contrivances which possessed principles now adopted in many aeroplanes.

The box-kite, for example, is used in a modified form in the tail construction of the Voisin, Farman, and Wright machines, as well as in several other aeroplanes. The Cody kite proved the sustaining-power of a given surface, and various other important points, and its inventor, profiting by his experiments, has had most noteworthy success with aeroplanes, one of which is the largest biplane ever built.

The tetrahedral kites of Alexander Graham Bell are most interesting structures, as they embody the strongest possible construction with the lightest weight and greatest area, and in their several odd and curious forms have proved most wonderful kites.

Briefly, the tetrahedral kite is a kite built up of triangular cells or compartments in various numbers and in any form. The simplest form of all is the triangular kite shown in

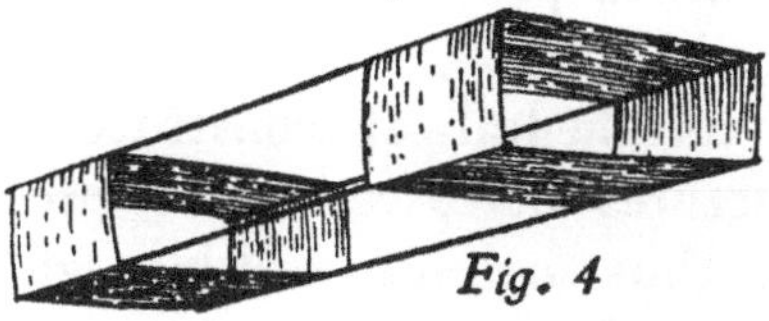

*Fig. 4*

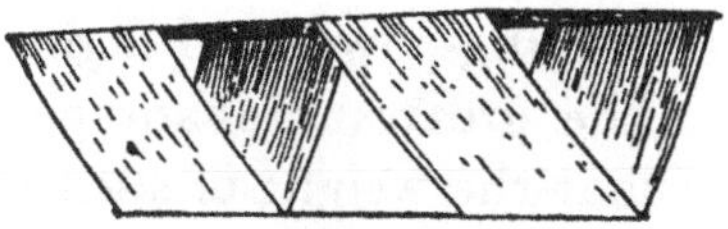

*Fig. 5*

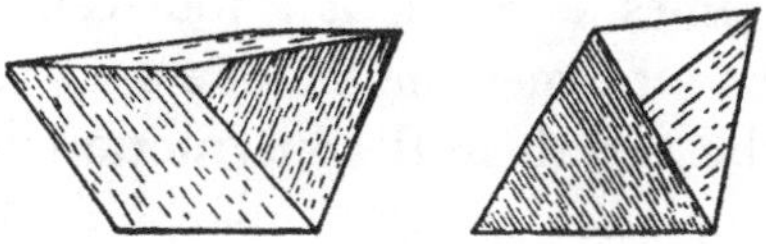


*Fig. 6*

A—A Triangular cell.
B—A winged tetrahedral cell.

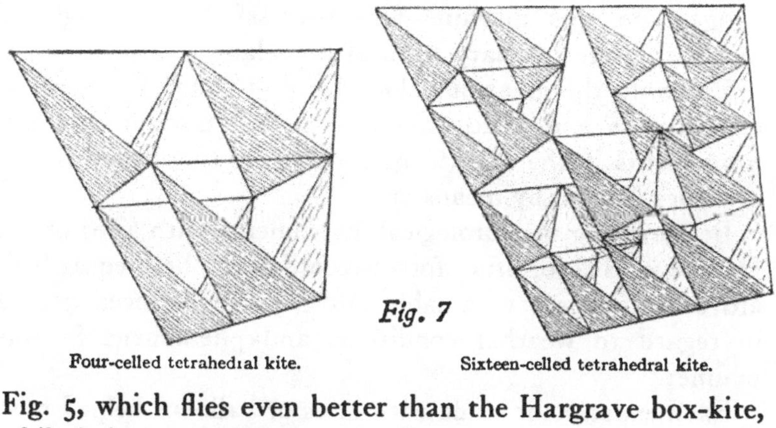

*Fig. 7*

Four-celled tetrahedral kite.     Sixteen-celled tetrahedral kite.

Fig. 5, which flies even better than the Hargrave box-kite, while it is many times stronger.

From these triangular compartments Mr. Bell developed the true tetrahedral cell, which is illustrated in Fig. 6, beside the plain triangular form.

By building up a number of such cells, each touching the next at its points, four, sixteen, and sixty-four celled kites were constructed, two of which are shown in Fig. 7. These kites were exceedingly efficient, and the same principle was used in forming round, rectangular, oval, and many irregularly shaped kites, all of which proved revelations in sustaining-power and flight.

The greatest advantage of tetrahedral-cell construction is that the weight of such structures increases in exact proportion to the area of surface; thus, a sixteen-cell kite will weigh but four times as much as a four-celled one, and will yet present just four times the latter's area to the air.

In practice the tetrahedral construction has not proved advisable in aeroplanes owing to mechanical difficulties involved and trouble in arranging reliable controls.

Quite recently an aeroplane has been built in which a bank of tetrahedral cells took the place of the ordinary wings, and this machine flew successfully, although no better than an ordinary form of aeroplane.

Probably the greatest value obtained from the countless experiments and discoveries in kite construction and improvements is the knowledge secured by the United States Weather Bureau by means of kites.

By elevating meteorological instruments with kites of the Hargrave, Bell, or other forms records are obtained at high altitudes, and most valuable information has been gained in regard to weather conditions and phenomena in this manner.

Each experiment made with kites or balloons helped toward solving the problems of flight, and vast numbers of experiments were carried on by numerous scientific men long before the Wrights even dreamed of flying.

Professor Langley constructed a model which actually

flew; Maxim built a huge aeroplane that destroyed itself and its supports by its lifting-power; and Chanute and others built practical gliders that, save for the motor, were very similar to our modern aeroplanes.

Although so very new in actual results, yet aviation is really very old, and many brilliant and earnest men have devoted their time, money, and brains for many years to attempts to conquer the air.

The Wright brothers made aeroplanes actually practical, it is true, but during the same period that they were carrying on their experiments in secret various American and European aviators were also accomplishing most astonishing results.

Santos Dumont actually flew in 1906; Farman in 1907 proved to all Europe that human flight was possible; and Voisin, Bleriot, and many others obtained equally noteworthy results.

The sudden advances made in aviation at that time and the ultimate success of the aeroplane were due, not to any one man or to any new discoveries in regard to aircraft, but to the perfection of the gasolene-engine.

The explosive motor solved the problem of human flight; it enabled us to produce submarines; it made possible the modern automobile and the motor-boat, and to the inventors of this wonderful compact, light, and powerful motor should be given the credit for our conquest of the air.

# INDEX

## INDEX



THE END

Spotted something that isn't quite right?

Let us know and we'll fix it for future reprints.
And as a thank you for your diligence, we'll
send you a free Sothis book of your choice.

publisher@sothis.press

This page intentionally left blank

Made in the USA
Coppell, TX
05 November 2026

11289008R10156